Quality Assurance
In Building

Macmillan Building and Surveying Series

Series Editor: Ivor H. Seeley
Emeritus Professor, Nottingham Polytechnic
Advanced Building Measurement, second edition, Ivor H. Seeley
Advanced Valuation Diane Butler and David Richmond
An Introduction to Building Services Christopher A. Howard
Applied Valuation Diane Butler
Asset Valuation Michael Rayner
Building Economics, third edition Ivor H. Seeley
Building Maintenance, second edition Ivor H. Seeley
Building Procurement Alan E. Turner
Building Quantities Explained, fourth edition Ivor H. Seeley
Building Surveys, Reports and Dilapidations Ivor H. Seeley
Building Technology, third edition Ivor H. Seeley
Civil Engineering Contract Administration and Control Ivor H. Seeley
Civil Engineering Quantities, fourth edition Ivor H. Seeley
Civil Engineering Specification, second edition Ivor H. Seeley
Computers and Quantity Surveyors A. J. Smith
Contract Planning and Contract Procedures B. Cooke
Contract Planning Case Studies B. Cooke
Environmental Science in Building, second edition R. McMullan
Housing Associations Helen Cope
Introduction to Valuation D. Richmond
Principles of Property Investment and Pricing W. D. Fraser
Quality Assurance in Building Alan Griffith
Quantity Surveying Practice Ivor H. Seeley
Structural Detailing P. Newton
Urban Land Economics and Public Policy, fourth edition P. N. Balchin, J. L. Kieve and G. H. Bull
Urban Renewal – Theory and Practice Chris Couch
1980 JCT Standard Form of Building Contract, second edition R. F. Fellows

Quality Assurance

in Building

Alan Griffith

M.Sc., Ph.D., MCIOB, FFB, MBIM
Department of Building, Engineering and Surveying
Heriot-Watt University
Edinburgh

MACMILLAN

First published 1990

Published by
MACMILLAN EDUCATION LTD
Houndmills, Basingstoke, Hampshire RG21 2XS
and London
Companies and representatives
throughout the world

Typeset by
Ponting–Green Publishing Services, London

Printed in Great Britain by
Billing & Sons Ltd, Worcester

British Library Cataloguing in Publication Data
Griffith, Alan
Quality assurance in building
1.   Construction industries. Quality assurance
I.   Title
624.0685
ISBN 0–333–52723–2
ISBN 0–333–52724–0 pbk

To Michela

Contents

Foreword		*xi*
Preface		*xiii*
Acknowledgements		*xv*

1	INTRODUCTION	1
1.1	The Need for Quality Assurance	1
1.2	A Framework for Quality Assurance	3
1.3	Application in Engineering Industries	4
1.4	Application in the Construction Industry	5
1.5	Application in Housebuilding	9
1.6	The Implications of Quality Assurance	10

2	QUALITY IN BUILDING	13
2.1	Definitions of Quality	13
2.2	The Meaning of Quality in Building	15
2.3	Quality Assurance in Building	17
2.4	Problems of Achieving Quality	20

3	QUALITY ASSURANCE STANDARDS	23
3.1	Philosophy and Development	23
3.2	The Structure for Quality Assurance in the UK	24
3.3	Sources of Quality Assurance Standards	27
3.4	BS 5750: Quality Systems – UK Standards for Quality	28
3.5	Quality Assurance: Types of Assessment	32

4	QUALITY ASSURANCE: CERTIFICATION	34
4.1	The Method of Third Party (full Certification	34
4.2	The Value of Quality Assurance Certification	41
4.3	Cost and Time Implications of Certification	42

4.4   Other Organisations Associated with Quality   47
      Assurance Certification

5   QUALITY ASSURANCE SYSTEMS   49

5.1   The Structure of Quality Systems   49
5.2   Guidelines to Quality Assurance Systems   54
      Development and Implementation
5.3   Detailing for Quality Manuals   58

6   QUALITY ASSURANCE:
    ITS APPLICATION IN CONSTRUCTION   66

6.1   Awareness within the Construction Industry   66
6.2   The Effect of Procurement System   68
6.3   Quality Assurance in 'Traditional'   69
      Building Procurement
6.4   Quality Assurance: Its Application in   87
      Non-Traditional Procurement
6.5   Quality Assurance in Building Services   94

7   QUALITY ASSURANCE IN HOUSEBUILDING   98

7.1   The National House Building Council (NHBC)   98
7.2   Development of the NHBC   99
7.3   The Maintenance of House Building Standards   101
7.4   The 'Buildmark' Scheme   102

8   QUALITY ASSURANCE: INTERNATIONAL INTEREST   105

8.1   European and Worldwide Dimension   105
8.2   Problems of Integration   106
8.3   International Quality Assurance and   107
      Certification Schemes
8.4   International Organisations Concerned with   116
      Quality Assurance

9   THE ACHIEVEMENT OF QUALITY: 119
    A REVIEW OF UK RESEARCH

9.1   Research Interest   119
9.2   Quality in General Building   120
9.3   Quality in Housebuilding   124

APPENDIX 1   DEFINITION OF TERMS   132

APPENDIX 2   SOURCES OF FURTHER INFORMATION   139
    Accreditation   139
    Certification Bodies   139
    Product Assessment and
     Certification Organisations   141
    Other Quality Assurance
     Related Organisations   141
    International Quality Assurance Organisations   143

Index   *145*

Foreword

Change has always been, and still is, one of the dominant features of the building process. The building industry is constantly faced with the challenges resulting from changes in design, materials, its production, technology, its management and even in the very way we think philosophically about the process of building. Perhaps the most significant aspect of change in recent times has been in the perception and understanding of the building process by the industry's clients. As clients today have a greater awareness for the building process and are no longer prepared to sit back and rely upon traditional procedures and sole guidance by the architect but rather become more actively involved in the process themselves, so greater demands are being made upon the building process to improve standards of performance, not only in terms of time, cost and quality, but also in project organisation, procurement and management. The building industry must respond to client's requirements quickly, efficiently and effectively, producing buildings that are inherently more buildable, which meet the client's genuine needs and which provide the best possible performance and value for money.

In recent times, growing attention has been given to establishing the Total Performance Concept for buildings. This allows clients the opportunity, from design brief through to commissioning and operating costs, to establish minimum standards of performance for all aspects of their buildings. This Concept offers an ideal framework within which to ensure all the necessary features of good quality.

The requirement for improved quality in building has never been greater and with this, a much greater emphasis is being given to developing managerial methods designed to ensure its accomplishment – Quality Assurance. The success of a building project relies, to a large extent, upon good practice in addition to good intent, that is the right attitudes bringing together the best skills. Quality Assurance requires appropriate systems, sound procedures, clear communication and documentation that is accurate and easily understood. These are the managerial prerequisites for Quality Assurance. Standards must be set and achieved. Quality Assurance impinges upon each and every aspect of the total building process and all building professionals should promote quality assurance as a common objective, indeed accept quality assurance as a duty to encourage others and produce themselves the highest possible quality throughout each stage of the building process.

The real benefits of Quality Assurance are only now beginning to emerge and be realised. The growing support for the adoption and development of formalised policies and procedures is well represented by BS 5750, the UK's national standard for quality systems and there is a strong trend towards the registration of quality systems (certification) throughout the UK building industry. Quality assurance does not stop there however. As Quality Assurance has emerged strongly in the UK in recent years so the European dimension has come sharply into focus. Recent developments in Quality Assurance management has seen BS 5750 become the foundation for the European Standard for Quality Assurance, EN 29000. Much is likely to follow from this milestone, not only within the domestic and European context, but also to put improved Quality Assurance firmly in perspective as an essential, challenging commitment for the building industry world-wide.

Professor V.B. Torrance
Past President
Chartered Institute of Building

Preface

In recent years there has been a dramatic upsurge of interest within the building industry for the greater promotion of Quality Assurance – managerial methods designed to plan, monitor and control the achievement of quality. The emphasis on Quality Assurance, thrust upon the building industry by BS 5750, the UK's national standard for Quality Assurance has pervaded the thought of all who procure, design and construct buildings. Quality Assurance within the building process is fast becoming a managerial application in its own right and which, to be successful, requires an understanding of quality and its management from the general to the specific.

This book sets out to provide a detailed guide to the principles and practices of quality assurance. Chapters 1 to 5 examines: the need for Quality Assurance in building; the meaning of quality; development of formalised methods or quality systems; and the approach to registration and certification following the requirements of BS 5750. Chapters 6 and 7 explore the application of Quality Assurance to the building industry and highlight: some of the problems that may be experienced; the roles of the various professional parties who can influence the achievement of quality; addresses the application of Quality Assurance to non-traditional forms of procurement and organisation such as project management and in other specific building sectors such as building services and housebuilding. Chapter 8 appreciates the importance of Quality Assurance in a wider context with cognisance being taken of the application of Quality Assurance in the international market through an insight into the national frameworks for Quality Assurance in Sweden, France, West Germany and the USA. Finally, the achievement of Quality Assurance can only be measured by its practical success. Chapter 9 reviews a number of research studies into the achievement of quality in the fields of general contracting and housebuilding and identifies problems that have occurred and actions required to seek improvement.

Throughout this book Quality Assurance is recognised as a fundamental requirement of the total building process. As such, it provides a valuable text for all students in the field of architecture, building and across the surveying professions and will also be of major interest to construction professionals in many sectors of the building industry.

Acknowledgements

The following professional bodies and institutions have, at some time, supported the background research upon which this work is based:

- The British Council and Fulbright Commission,
  Higher Education Division

- The Chartered Institute of Building,
  Queen Elizabeth II Silver Jubilee Scholarship Fund

- The Fellowship of Engineering,
  Research Fellowship Awards

- The Carnegie Trust for the Universities of Scotland,
  Higher Research Scholarship Fund

I am also grateful to Professor Ivor Seeley, the Series Editor, for reading through and commenting on the draft typescript.

1 Introduction

1.1 The Need For Quality Assurance

In recent years, increasing concern has been expressed at the standards of performance and quality achieved in UK building work. The need for structured and formal systems of construction management to address the aspects of performance, workmanship and quality has arisen as a direct result of deficiencies and problems in design, construction, materials and components. Many of the problems experienced in building appear as a range of inadequacies from minor technical and aesthetic aspects to major building defects. Irrespective of their degree of severity, such problems are known to cost the industry hundreds of millions of pounds annually, yet, many difficulties might be alleviated through greater care and attention to standards of performance and quality at the briefing, design and construction stages of the building process.

The focus of blame for inadequate building performance and inferior quality is frequently appropriated to the architect. It is not uncommon for the architect to be severely criticised for: a lack of awareness for the building process on site; or for failing to understand new technology and the performance of innovative materials; or the reluctance to delegate authority to project based supervisors; or for simply not spending more time on site. Whilst some of these criticisms may, indeed, be justified, it is not always appropriate and not just the architect who may be at fault.

Building clients may be found lacking in specifying clearly their desires and requirements for a building and frequently they have difficulty in deciding quite what they want, in terms of building performance and standards of quality. Building contractors could spend more time structuring a planned approach and organisation towards meeting the requirements of quality rather than progress chasing. They could also give greater attention to leading and motivating the workforce towards achieving better workmanship.

Forms of building procurement and contractual systems are sometimes pinpointed as being responsible for inadequate standards of project performance and building quality. The 'traditional' procurement approach, where each building professional acts in an independent role, structures a project organisation in which, it appears, a main aim of the contractual parties is to apportion blame for project inadequacies, rather than integrating their resources and effort towards achieving project success.

There should be changes in the current role, duties and responsibilities of individuals and organisations within the building process. There should be changes to procurement systems and forms of building contract, but perhaps more important, in the immediate term, there should be greater awareness developed for the needs of improved performance and quality within the total building process. Moreover, there should be a change in the attitude towards interpreting quality within the construction industry. 'Quality Assurance' the managerial concept of devising methods by which quality can be planned, monitored and controlled, should not be perceived as a burden, but rather accepted as a challenge and a commitment.

Change is being brought about, albeit slowly and not always for the right reasons. The importance of quality and quality assurance to a designer, contractor, supplier or manufacturer, is sometimes not a question of integrity and the intrinsic desire to provide a quality product or service, but more a requirement to maintain or increase their commercial share of available work or indeed, their involvement in a fundamental fight for survival in the market place.

The need for quality assurance is an implication of the many inefficiencies and problems in buildings, the consequential occurrence of protracted litigation and a loss of professional, commercial and public acceptance of building. Many public and private sector clients are becoming increasingly dissatisfied with the difficult operational characteristics of the traditional form of building procurement. Clients are becoming more aware of the workings of the construction industry, have a greater appreciation for building performance and quality and actively seek greater involvement in the design and construction processes. Today, clients demand more novel, non-traditional, procurement systems which place a greater emphasis upon providing performance, quality and better value for money.

The public also are adopting a more discerning attitude in the choice of building product or service they purchase. They too are seeking more reliable ways to engage the services they require from the construction industry and this itself is beginning to eliminate poor services and invoke greater professionalism within the building process. Such demand-pull influences are always slow to emerge but a beneficial trend is evident nonetheless. In particular sectors of the construction industry, such as housebuilding, the changing requirement of the public is clearly evident. The housebuyer today holds ever increasing expectations for building performance and quality in their new homes and housebuilders are forced to respond directly to this demand to maintain their commercial position and standing.

Building clients and customers are taking a far more intuitive stance in their acceptance of quality, in particular with regard to building products and materials. The increasing liability manufacturers face from goods not meeting their intended fitness for purpose is challenging the very basis of their business. The sometimes apparent ignorance or commercial disregard for quality is being replaced by a structured approach to quality assurance. There is far greater emphasis upon 'preventive' quality assurance rather than a mere quality control acceptance–rejection approach.

The emphasis upon quality assurance, thrust upon the construction industry by BS 5750: Quality Systems, has made everyone who designs, constructs, installs goods, or supplies services, more aware of their onerous responsibility for quality and performance. The implementation of a formalised Quality Assurance System, to the requirements of BS 5750, does not guarantee quality but serves to ensure as much as possible that a high quality product or service is provided such that the risk of deficiency or failure is minimised.

Quality assurance to the client or purchaser is a test of the commitment and pursuit of quality to which the product or service has been subjected. It reduces the risk taken in relying on the success of his purchase. To the customer, quality assurance is a cost effective means of obtaining a product of known quality, recognised performance, and above all, a purchase that represents better value for money. To the manufacturer, designer or building contractor, quality assurance is a way of demonstrating confidence in the product or service being marketed to industry and highlights the importance and commitment given to the pursuit of better quality and value. Quality assurance is not always a cost effective activity but is essential if fitness for purpose is the measure of performance and where satisfaction of the client or customer is to be placed first and foremost.

1.2 A Framework For Quality Assurance

'Quality Assurance' which encompasses 'all activities and functions concerned with the achievement of quality' (BS 4778), has only come to the fore within the construction industry during the 1980s, although concern for quality dates back much earlier in other sectors of industry, most noticeably in manufacturing and engineering.

The origins of quality assurance are clearly in manufacturing industry where they are recognised in the procedures of quality control. Quality control however, is essentially a matter of inspection and checking the performance of a product output against a predetermined standard and rejecting those items which fail to conform.

This analogy is somewhat simplistic and, of course, it was recognised early in the development of quality management in manufacturing that acceptance–rejection testing is inherently wasteful and that quality must be 'manufactured-in' to the product. Likewise, it was realised that there is little use in comparing the output to a specification if that specification has no element of quality 'designed-in' to its concept. Thus, quality soon became a concern for design, manufacture and use, and led the way for a more structured approach towards quality assurance.

Quality assurance is well established in the engineering industry and to a much lesser extent in building, although the standard used as a basis for assessment, BS 5750, has its roots in manufacturing industry. Quality assurance became prominent in the 1960s as a management concept used in the aerospace, nuclear power industries and Governmental defence programmes of the USA and was first used in the UK to meet similar requirements. Early implementation of quality standards such as BS 4891 have been superseded, since 1979 by BS 5750, the UK's national standard for quality assurance systems, which itself was revised in 1987.

BS 5750: Quality Systems is presented in six parts. Parts 1 to 3 state the requirements for three areas of quality assurance systems:

Part 1: Specification for design/development, procedure, installation and servicing.
Part 2: Specification for production and installation.
Part 3: Specification for final inspection and test.

Parts 4, 5 and 6 present guidance to the practical implementation of quality assurance systems specified in Parts 1, 2 and 3.

Since BS 5750 has its origins in manufacturing industry, its requirements need some interpretation when applied to use in the construction industry. Formal quality assurance systems to the guidelines of BS 5750 suggest that the manufacturer of materials and products or supplier of a service to industry formulates, develops and implements a workable, organised and documented procedure for ensuring the quality of the product or service. Quality assurance in its full form, or 'third-party' quality assurance systems, measure and assess that system against the national framework for quality systems – BS 5750 and where compliance is complete the system can be registered (certification) as a recognised mark of achievement and excellence to the pursuit of quality.

1.3 Application in Engineering Industries

Quality assurance has, for many years, been a characteristic of the engineering industries, UK nationalised industries and the larger private sector companies, many of whom have developed their own standards for performance and quality within their particular industry sectors. Such industries have developed further the application of quality assurance by carrying out the assessment of companies to establish published lists of approved and reputable suppliers and services.

Within process engineering and manufacturing, the expected standards of performance and quality, in both design and assembly, are clearly determined before the work is undertaken with all designers, contractors, sub-contractors and suppliers being quality assessed as a condition of their appointment. Not only is quality assurance a prerequisite to procurement, but quality is continually appraised through monitoring and inspection during manufacture, delivery and assembly.

Quality assurance in engineering is very formalised and thorough to ensure that all materials and products meet their fitness for purpose and that all services meet their contractual obligations. The complete range of quality assurance services in engineering will meet requirements in the following areas:

- Quality systems development
- Quality systems organisation and procedures
- Quality systems inspection and testing
- Quality systems auditing
- Quality systems review

This is the general pattern for quality assurance development that has been followed in recent years by the construction industry.

1.4 Application in the Construction Industry

Within the construction industry the application of quality assurance has been slow to emerge. Whilst construction must formally meet with requirements specified by the Building Regulations and conform to planning and building approvals set by local authority building control, there is no compulsory requirement for design practices, manufacturers, contractors, or consultants to conform to the requirements of BS 5750: Quality Systems. BS 5750: Quality Systems, is at present, advisory in nature and its adoption is a voluntary undertaking, influenced to a large extent, upon the philosophical or commercial interests of the company.

With the introduction of BS 5750, there has been considerable disparity of opinion over what quality assurance seeks to achieve. There is much confusion between the aims of quality control and quality assurance and where and when the principles of quality assurance should be applied within the building process. As the principles of quality assurance are not widely understood nor accurately interpreted within construction, it is often argued that the application of quality assurance in construction is intrinsically difficult. It has been said that quality assurance can only be applied under strictly controlled environmental conditions such as those that prevail in manufacturing, and as such, its principles have little or no relevance to building with its complex and fragmented nature and where in practice work patterns can change from day to day. Despite such opinion, quality assurance has found considerable support and application in a number of significant areas within the construction industry.

It is paramount that BS 5750: Quality Systems be treated with considerable respect. BS 5750 has been severely criticised in recent years. Major concern surrounds the adoption of principles from manufacturing industry and their application to the distinctive nature of building. It may be the case that the fundamental nature of building and the methods of procurement and organisation used, do in fact serve, upon occasion, to impede rather than enhance quality in building today. The unpublished report *A Strategy for Quality Management Systems in the Construction Industry* (1987) puts forward a number of major characteristics of construction work which differ from the approach of BS 5750 and therefore challenge the validity of the standard when applied to the building industry. There is, however, sufficient similarity between quality management in manufacturing and particular sectors of construction to enable the standard to be used as a guideline but, like everything else which requires interpretation in application, it must be used wisely. It is, therefore, essential to emphasise that BS 5750 requires considerable and careful interpretation in its application to building. It is primarily aimed at providing a systematic approach towards the better management of quality in building. The nature of the building process itself and the application of standards such as BS 5750 can never guarantee quality. Used sensibly however, BS 5750 can provide an accent towards achieving better quality. Whilst being far from ideal, BS 5750 does present a general framework for quality assurance systems within which any organisation can develop its own system to meet its own needs. Quality assurance in building is concerned explicitly with systematically ensuring compliance with the project requirements, providing a working system that assures work is, where possible, to the highest standards achievable, and encouraging an attitude that quality within the building process exists to be managed.

Outwith the 'guise' of quality assurance, it should be emphasised that 'quality management' has been the norm for many years. Any large organisation carrying out major projects will have, through its structure, organisation and procedures, invoked quality assurance as an intrinsic if subconscious activity. Quality assurance could be said therefore to be little more than good general management practice. The distinguishing characteristic of quality assurance is in its recognition for structure, formality and that management techniques and activities addressing quality assurance stand out from the general practice of organisational management, that is, it becomes a managerial concept in its own right.

There are a number of third-party quality assurance schemes currently in operation with specific relevance to the building and construction industry. The most commonly known schemes are the British Standards Institution (BSI) 'Kitemark' scheme and the certification scheme of the British Board of Agrement (BBA). These schemes are essentially product conformity and approval mechanisms. Quality assurance in respect of construction, management or consultancy services to the building industry comes within the remit of the British Standards Institution: Quality Assurance Services scheme (BSI/QAS).

There are also several independent quality assurance schemes. These cover product conformity assessment such as the 'Testguard' certification of products provided by Yarsley Quality Assured Firms Ltd, formally the Yarsley Technical Centre. Independent services also provides quality assurance certification schemes for management and construction inputs to the construction industry, the most relevant certification bodies being Yarsley Quality Assured Firms Ltd, (YQAF), and Lloyds Register for Quality Assurance, (LRQA). Such certification bodies and their schemes are accredited, or approved by Government, through the Department of Trade and Industry (DTI) and the DTI's functioning body in this field, The National Accreditation Council for Certification Bodies (NACCB). In this way, a framework for quality assurance has become fully developed and operational in the UK. Figure 1.1 illustrates the overall structure.

The BSI Kitemark awarded to manufacturers assessed, approved and licensed by BSI indicates that samples of their product have been independently tested against the appropriate British Standards, and certifies that the product meets with those standards. The British Board of Agrement tests, assesses and certifies new products or the new and innovative use of existing products. The BSI/QAS and the independent certification services provided by Yarsley and Lloyds develop, monitor, assess and certify that a company employs an effective quality assurance system within their organisation and field of operation. Schemes such as these therefore, present a capability for quality assurance throughout the building process.

With all certification schemes it is important to appreciate that assessment of a quality assurance system is based upon 'the observation' of an organisation's overall implementation of management and production procedures, for example a periodic cross-sectional view of the quality system. These observations may not always represent a true picture of the day to day application of quality assurance within the company. The implementation of a quality system does not, therefore, guarantee or insure against inadequacy or failure within construction, but provides only an assessment of the organisation's ability to work to high standards of quality. Insurance protection against building failure as a result of deficient materials supplied or wrongly specified materials, therefore, remains important to designers

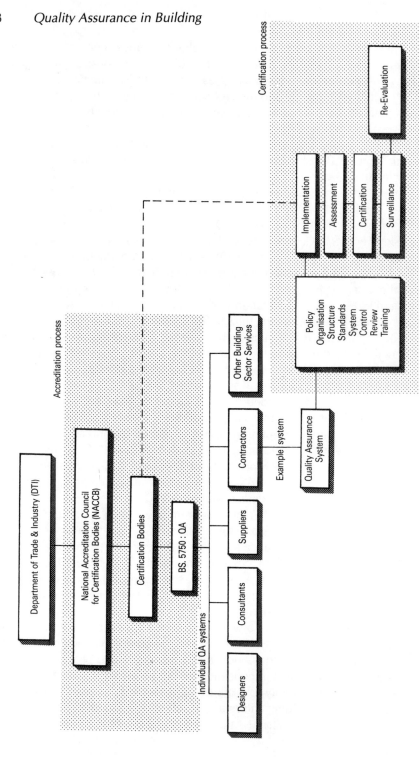

Figure 1.1 Framework for quality assurance in the UK

and manufacturers. The only sector of UK construction industry that provides the purchaser with any guarantee of performance or quality backed by insurance is that of housebuilding.

1.5 Application in Housebuilding

One of the most successful applications of independent third-party quality assurance certification can be seen in UK housebuilding. The National House Building Council (NHBC) is an independent non-profit making, non-political body, financed through levies paid by individual house-builders. The NHBC sets standards for new homes in addition to those requirements specified by building regulations and local authority planning and building control. The overall aims of the Council are to set recognised standards for all new houses and to guarantee the design and construction of new homes against major building deficiency and defect. It protects the purchaser from inadequacies in housebuilding performance and quality in a number of ways:

- Setting minimum standards of construction to which housebuilders must comply.
- Inspecting houses during construction.
- Providing a ten-year warranty on the completed house.

Almost all private housebuilders in the UK are registered with the NHBC. The principal reason for the virtually total acceptance of NHBC specifications is that unless a new house has been NHBC certified for standards of quality and workmanship, it is extremely difficult, if not impossible, to finance its purchase since the scheme is well supported by the major building societies and banks. The NHBC certificate covering a new house is a pre-requisite of securing a mortgage and forms a condition of contract between the financier and the purchaser.

The NHBC certified houses are guaranteed under a two-year warranty, backed by insurance, against defect from work which complies with the specification set by NHBC. There is, in addition, a further ten-year guarantee covering major structural defects and also an insurance protecting the purchaser against the housebuilder going into liquidation and failing to meet his obligations under the warranty.

The monopoly of NHBC has, in recent years, been challenged with the introduction of an alternative housebuilding warranty, a fifteen-year scheme, 'Foundation 15', provided by the Municipal Mutual Insurance Company (MMI) an independent housing insurer. This competition has been useful in stimulating improvement to the NHBC scheme. Since April 1988, the ten-year warranty scheme has been improved through the introduction of 'Buildmark', which extends the nature and scope of NHBC cover beyond the original standards. The NHBC warranty scheme remains the construction industry's foremost application of quality assurance at this time.

1.6 The Implications of Quality Assurance

The application of quality assurance within the construction industry is fraught with difficulty. The relatively unique aspect of each construction project and its temporary state make the continuity and development of approach across projects far from easy. In addition, quality assurance must address, not only the science and technological aspects of construction, but manage the physical and psychological aspects of the human element. Quality assurance in construction therefore, represents a complex socio-technical managerial challenge.

Within the traditional form of building procurement, quality assurance relies upon the individual contribution to implementation from each designer, contractor, supplier and sub-contractor. Given that each professional party ostensibly acts in isolation, the aims and objectives of quality assurance are easily compromised and frequently lost. The adoption of non-traditional procurement practices can, however, make a useful contribution to quality improvement.

Quality assurance is firmly dependent upon, clients knowing their specific needs and communicating these unambiguously to the designer, upon the designer accurately representing these requirements in the design concept, upon the contractor faithfully reproducing these requirements in the work on site, and taking quality assurance to its end, upon the occupier using the building correctly to achieve maximum performance.

It is unfortunate in building that product specification or procurement of services is not always based upon its potential for performance and quality but selected erroneously upon speed of delivery and assembly or upon cost. Similarly the quality of manufactured products or materials cannot be guaranteed as cost may have been a more significant criteria than inherent quality in production. Manufacturers and service companies may need to minimise their overheads simply to survive in the marketplace. The strategic significance of quality in marketing and the balance of quality with cost is well appreciated, but there is far less awareness for 'quality cost', the somewhat hidden costs associated with remedial work resulting from defective practices. In the construction industry such costs are difficult to determine but are known to run into millions of pounds annually. This situation is somewhat accepted within UK construction industry. Little is made of the fact that the most significant proportion of total quality costs is associated with 'failure cost', (the cost of reject work and claims for rectification), whilst a relatively minuscule proportion is given to 'preventive cost', (the cost incurred to eliminate or reduce the causes of quality failures).

The successful implementation of quality assurance can change the implications of quality cost fundamentally, since the primary aim of quality assurance is to systematically 'prevent' deficient performance and inadequate quality. An organisation in any sector of construction industry which believes that the implementation of quality assurance will result in it pricing itself out of the commercial marketplace or that it unacceptably increases the cost of its operation, does not understand, or has no wish to understand, the real contribution of quality assurance. Achieving the right standard of performance as required by the client or customer and presenting high quality service or products is simply good for business. The 'right' standard is that which will maximise the turnover and long-term profit of the organisation. The right standard is not the cheapest or simplest standard to achieve in the immediate term, nor that standard which is invoked with reluctance.

The UK construction industry is some way yet from the situation where all architectural practices, consultants, suppliers and contractors have a recognised quality assurance system and third-party certification becomes a mandatory requirement. The construction marketplace is however, beginning to become more aware of the need for quality assurance and thus has a growing following. Already, a number of public sector and major private sector clients are demanding thorough proof of a contractor's credentials for quality assurance capability before compiling their tender lists. Following the present trends, the future may see quality assurance systems becoming an inherent aspect of building contracts whereupon, quality assurance will be a compulsory requirement imposed by industry itself.

Many of the larger and most respected building contractors have, in recent years, developed in-house quality assurance systems but few have, as yet, perceived a critical need to extend their systems to full third-party certification. This requirement may, indeed, be a few years away and many contractors will wish to restrain potentially costly reorganisation until the benefits and rewards of implementing quality assurance are overwhelmingly present. It is unfortunate that a number of research studies, undertaken since the late 1970s in the area of housebuilding and general building work, have reported some deficiencies in building performance and quality. These studies have illustrated unequivocally that existing problems of quality assurance are too common and remain unresolved. The construction industry requires clear evidence of the benefits of quality assurance if support and implementation is to self-perpetuate. This requirement is simply lacking at this time, and undermining the real potential of quality assurance.

Although quality assurance has been slow to emerge and develop in the UK construction industry for many reasons, it is clear that quality assurance is the way to go. Awareness must become more widespread, its theories appreciated, ideas tried, and its concepts developed. It is also clear that the construction industry cannot rely upon its traditional procedures and customs and upon the ad hoc approaches to quality management it has supported in the past. Quality assurance is the concern of all. No sector of the construction industry should consider itself isolated or excused from its application. Everyone within the construction industry has a duty of care to ensure that quality assurance becomes a fundamental commitment and achievable goal.

Fundamental changes in the technology and structure of the building industry mean that a far more systematic approach to quality is required of everyone involved in the design and construction of a building.

Sir Monty Finniston
Chairman, Building Economic Development Council
[Achieving Quality on Building Sites,
NEDO Report (1987)]

2 Quality in Building

2.1 Definitions of Quality

Quality

Definitions of quality abound. For many years there have been attempts to define the meaning of quality, often in general terms, yet more recently in terms of the formalisation of quality through 'Quality Assurance Systems'. Some definitions result from authoritative documentation whilst others express experiences, opinions and conjecture. Although considerable disparity pervades, there is also much commonality amongst various definitions.

The definition of quality in British Standard – BS 4778 (1971) [1] acts as a reference point:

> The totality of features and characteristics of a product or service that bear upon its ability to satisfy a given need.

The Building Research Establishment (BRE) Report, 'A Survey of Quality and Value in Building (1978)', [2], provides further explanation in the context of construction:

> The totality of the attributes of a building that enable it to satisfy needs, including the way in which individual attributes (external attributes; performance attributes; and aesthetic and amenity attributes) are related, balanced and integrated in the whole building and its surroundings.

These definitions, widely recognised in the construction industry, suggest quality is interpreted as a 'fitness for purpose' (Construction Industry Research and Information Association (CIRIA) Report No. 109, (1985) [3]. This interpretation attempts to make quality finite, factual and measurable, yet quality in practice remains quite subjective in nature. Quality is not perfection, but provides some measure of standard or requirement.

13

Quality Assurance

Quality Assurance (QA) encompasses:

All activities and functions concerned with the attainment of quality. (BS 4778.)

Quality Assurance is:

a systematic way of ensuring that organised activities happen in the way they are planned. It is a management discipline concerned with anticipating problems and creating the attributes and controls which prevent problems arising. (CIRIA Report No. 109 (1985).)

The overall aim of quality assurance within the construction process is to provide the client with a product that will be:

- Suitable for the intended purpose
- Property constructed
- Satisfactory in performance
- Value for money.

Quality Assurance is therefore:

an objective demonstration of the builder's ability to produce building work in a cost effective way to meet the customer's requirements. (Chartered Institute of Building (CIOB) Report: Quality Assurance in Building (1987).) [4]

Quality Assurance also involves:

a management process designed to give confidence to the client by consistently meeting stated objectives. (Royal Institution of Chartered Surveyors (RICS), Quality Assurance: Introductory Guidance (1989).) [5]

CIRIA Report No. 109 describes quality assurance more fundamentally as:

simple common sense written down.

Quality Assurance is therefore, essentially a product of good management practice. It should not be an over-elaborate management technique but rather a co-ordinating and recording mechanism that minimises the risk of inadequate performance at all stages of the construction process.

Quality Management

To provide a good quality product, there must be unambiguous management objectives and organisation, or a 'Quality Management' (QM) System, which places emphasis upon the achievement of quality.

A Quality Management System is defined in BS 4778 as:

The organisation structure, responsibilities, activities, resources and events that together provide organised procedure and methods of implementation to ensure the capability of the organisation to meet quality requirements.

In BS 5750, this Quality Management definition is developed as follows:

The organisation structure, responsibilities, activities, resources and events appertaining to a firm that together provide organised procedures and methods of implementation to ensure the capability of the firm to meet quality requirements established in accordance with Part 1, 2 or 3 of BS 5750. [6]

Quality Management, in simple terms, is therefore:

That aspect of the overall management function that determines and implements the Quality Policy. [5]

Quality Management Systems and Quality Assurance should be an instinctive aspect of the building process. The search for quality is:

concerned with developing and planning the necessary technical and management competence to achieve desired results. It is also about attitudes, both of management and of all those for whom they are responsible. The philosophy underlying quality assurance is intended to ensure the provision of building work which satisfies the customer's requirements and offers a fair return for the resources employed. As such, it becomes a way of doing things. [4]

2.2 The Meaning of Quality in Building

Understanding Within Construction

Formal Quality Assurance systems to aid management have, for many years, been characteristic of manufacturing industry but are somewhat evolutionary within construction practice. Although analogous in part, the design and production of a building differs significantly in many ways from the design and manufacture of products.

It is important to appreciate some essential differences with regard to quality in building.

- Almost all construction projects are 'unique' with the building process representing a single production run.
- Tradition tends to separate the design and construction processes, whilst manufacturing industry adopts integrated processes.
- The construction site is 'individual' in terms of its temporary environment.

- The life cycle of a construction project, from inception to completion, extends beyond the manufacturing cycle and also tends to evolve and develop through time.
- Considerable mobility of design and construction staff preclude development of long-term production teams and each construction site is likely to have different team members.
- There are, as yet, no 'precise' standards specified for design and construction quality whereas in manufacturing industry there are carefully controlled tolerances.
- Supervision and inspection of construction work is less systematic than in manufacturing where inspection procedures can be clearly regulated.
- Feedback from the building in use to the designer is remote from the actual time of design and construction and often precludes the effective analysis of defective design and construction, whereas in manufacturing testing for deficiency and necessary remedial action can be implemented quickly.

The management of quality in the design and construction processes is being directed to 'ensuring' quality, that is, the provision of formal quality assurance. But, what does it mean to introduce quality assurance into the design and construction processes? Essentially, an assurance is being sought that the design and construction aspects have the capability to produce a product that is effective, efficient and economic, whether that product is the design of the building or the construction of the building. Moreover, the pursuit of quality commences with the client and continues through the production process to the building in use. Quality Assurance, therefore, becomes an integral part of the 'total building process'.

The Expectations for Quality

Although 'quality' can be defined, the perception, interpretation and measurement of quality lacks clear description. Superficial perceptions have described quality as differences in structural stability, precision, durability, and more frequently as appearance. Quality is, in many ways, subjective and, therefore, becomes a matter of judgement. For a clear perception there are a number of aspects which should be considered. These aspects are:

(i) *Function*   does the building meet the requirements?
(ii) *Life*   is the building durable?
(iii) *Economy*   does the building represent value for money?
(iv) *Aesthetics*   is the building pleasing in appearance and compatible with its surroundings?

(v) *Depreciation*   is the building an investment?

The interpretation and measurement of quality are equally as ambiguous as its perception. 'Notional Quality' describes the client's needs and desires for a quality product. The architect's pursuit of value for money, often assisted by a quantity surveyor, aims to provide an acceptable standard of construction, to a respectable cost and produced in a feasible production time. At the workplace quality is directed towards the skill and application of the craft operative, or 'workmanship'. Interpretation, therefore, depends fundamentally upon one's standpoint within the construction process.

Problems surrounding perception, interpretation and assessment result from the lack of measurement techniques, yardsticks and conclusive tests to determine levels of construction quality achieved. Whilst building materials, products and components are marshalled to a large extent by regulation and control standards, no standardised performance schedules exist to measure the quality of the human resource. Quality systems which cover a variety of types of system for design, manufacture, installation and testing are described in BS 5750, but there is no specific reference to construction practice. Quality in construction is, therefore, frequently deter- mined by expectation. Dalton [7] emphasises that the management of quality and quality itself are closely related to a number of discernible expectations surrounding the performance of buildings, these being quality, durability and reliability. Quality is emphasised as this represents the measure of fitness for purpose as defined in the client's brief. Durability forms a measure of how well a building and its component parts have the ability to withstand wear and to meet the client's long-term expectations. Reliability is essential as this is a measure of how well and consistently the building(s) and its parts perform with reliability being judged by the things that go wrong, the way they go wrong and how often they go wrong.

2.3 Quality Assurance in Building

Quality Assurance is concerned with developing a 'formal' structure, organ- isation and operational procedure to ensure good quality throughout the total building process. 'Quality' is generally used as a measure of the fitness for purpose, in the sense of meeting the needs of the client and 'Assurance' comes from the assessment and recognition of an organisation's quality management system by an independent assessor – 'the certification body'.

Quality Assurance is not a concept for application across the construction industry per se, but develops an industry wide framework within which, individual companies develop their own quality systems to the guidelines of BS 5750: Quality Systems, the national standard by which Quality Assurance Systems are currently assessed.

Building designers and contractors, unlike manufacturers, may be involved in a wide variety of projects. Quality Assurance within the construction industry must, therefore, give clients confidence in the ability of designers, contractors, suppliers and other parties alike to address and meet their particular requirements. Quality Assurance certification must include an assessment of the scope of experience and efficiency in relation to management and the technical and financial competence of all parties involved in the building process. Whilst quality assurance and management systems differ in nature and format and in the various sectors of construction, their principles and direction are singular in one aspect – the pursuit of 'better quality' in building.

Quality Assurance and the Total Building Process

Within the construction industry there are five broad sectors where quality assurance is applicable:

- (i)   *Client*   in the project brief.
- (ii) *Designer*   in the design and specification.
- (iii) *Manufacturers*   in the supply of materials, products and components.
- (iv) *Contractors (and sub-contractors)*   in construction, supervision and management.
- (v) *User*   in the use of the new building, in its upkeep and repair.

There are few building standards and codes of practice in the client area of building procurement and similarly few concerning the occupation and use of the finished product. The vast majority of quality assurance applications are in the manufacturing sector of the construction industry where factory production with systemised quality assurance systems ensure uniformity of product standard.

BS 5750: Quality Systems, presents a framework for the development of quality assurance in all five sectors of the construction industry. Any system of quality assurance must encompass the following:

- (i)   There must be a 'formalised' quality assurance system and organisation.
- (ii) The system must be implemented.
- (iii) The system must be 'recognised', hold current certification and be monitored/enforced by an independent body.

These aspects are guided by BS 5750 to which all quality assurance systems must comply and operate.

The final quality of a new building, its materials and its component parts will depend upon a number of factors. These are outlined by Fletcher and Scivyer [8]:

(i) That the product standard and specification properly defines the fitness for purpose.
(ii) That the product accurately represents the specification.
(iii) That the product is delivered and stored on site so as to avoid damage and hence reduced performance.
(iv) That the product is correctly installed, for example:
  a) That the design of the building makes it explicit how the product is to be combined with others as part of the building to give effective performance.
  b) Installation is carried out properly in accordance with the designer's intentions.

Whilst the first two factors are reflected in quality assurance schemes for manufactured products, the remaining factors address much wider issues which have greater influence upon quality in construction as these introduce the management, organisation and operational aspects of construction.

In order to 'manage' quality, each stage of the procurement process through which the construction project evolves must be defined, and their influences upon quality determined. Hill [9] specifies four categories.

(i)   Quality of the design process
(ii) Quality of the construction process
(iii) Quality of the products
(iv) Quality of maintenance

Quality of the design process includes reliability of the initial brief, reliability of the design solution and the detailed specification, reliability of all the information that has been used as the basis for the design and product specification.

Quality of the construction process requires the reliability of the organisation, procedures, and skills of the builder to interpret the design and provide the end product on the site in accordance with the design/specification.

Quality of the products needs reliability in all the materials/products/components incorporated in the building.

Quality of maintenance is the reliability of the upkeep, maintenance and repair programme and that the building use is not modified in such a way as to significantly affect its performance.

These requirements place responsibility upon the five main parties involved in construction: client, designer, manufacturers, contractor, and user. For effective quality assurance, an overall quality system must be traceable across the total building process and the individual company quality assurance systems be implemented and the interfaces between individual systems be tightly controlled. The onus is clearly upon all parties involved in the building process to recognise BS 5750 as the conceptual framework and structure for promoting 'their' strategy and efforts towards improving, not only their own excellence but also through the contribution towards improving the 'quality of building'.

2.4 Problems of Achieving Quality

The vast majority of building failures are thought to occur from faults originating in the quality of design and the construction process. Irrespective of their nature, such faults are due to a lack of management or to the quality of management. These can range from minor problems such as 'snagging items' whilst others can be latent defects resulting in protracted litigation. These problems pervade due to:

(i)   Historic separation of design from construction.
(ii) Poor communication of design requirements.
(iii) Design is difficult to build.
(iv) Poor labour skills and supervision.
(v)   Complex contract documentation.
(vi) Unrealistic time and cost assessments.

Specific problems surrounding the achievement of quality can be broadly categorised into two areas, problems attributable to:

(1) Design:
  (a) *Detailing* – inaccurate or inadequate detail of design concepts.
  (b) *Specification* – incorrectly specified or misused materials and components.
  (c) *Legislation* – inadequate knowledge or disregard for compulsory legislation or advisory documentation.
  (d) *Co-ordination* – inadequate co-ordination between client/designer and contractor.
  (e) *Communication* – poor interaction between client and designer; and designer and contractor.
  (f) *Supervision* – inadequate supervision of construction by client and designer.
  (g) *Buildability* – lack of design empathy for construction.

(2) Construction:
  (a) *Project Priorities* – speed and cost factors outweigh the requirements for quality and workmanship.
  (b) *Organisation* – inadequate definition of site duties to the workforce.
  (c) *Information* – yardsticks of performance are not prescribed and information flow is poor.

(d) *Control* – few quality control procedures used at site level.
(e) *Supervision* – inadequate site management and foremanship.
(f) *Workmanship* – inadequate standards of work at the workplace.
(g) *Motivation* – inadequate skilled and motivated operatives.
(h) *Co-ordination* – lack of teamwork and direction.

The Value of Quality Assurance

Quality Assurance Systems must be based upon developing a design and construction process that assures quality in the design, in the materials and components used, in the site procedures used for communicating performance standards from design to the workplace, and in effective monitoring, control and feedback mechanisms.

Effective quality assurance will lead to advantages, both for the design process and construction phase. The implementation of formal quality assurance procedures can bring:

• Better design
• More effective planning
• Improved site management
• Increased project performance
• Efficient management of construction problems
• Improved quality
• Fewer delays and disruptions
• Lower cost of remedial and repeat works
• Provision of feedback for future projects
• Enhanced reputation for good design and construction

The emphasis in construction industry today, is therefore the formulation, development and implementation of 'Quality Assurance Systems'.

References

(1) British Standards Institution *(BSI) BS 4778 (1971)*
(2) Building Research Establishment (BRE) *BRE Report: A Survey of Quality and Value in Building (1978)*
(3) Construction Industry Research and Information Association (CIRIA) *CIRIA Report 109 Quality Assurance in Civil Engineering (1985)*
(4) Chartered Institute of Building (CIOB) *Quality Assurance in Building (1987)*
(5) Royal Institution of Chartered Surveyors (RICS) *Quality Assurance: Introductory Guidance (1989)*
(6) British Standards Institution (BSI) *BS 5750: Quality Systems (1979)* revised 1987

(7) Dalton J.B. Property Services Agency (PSA) *Quest for Quality: Developments in the Management of Quality by the United Kingdom Department of the Environment's Property Services Agency (1988)* CIB W/65 Organisation & Management Construction Proceedings Volume 1, pp. 360–366, Spon

(8) Fletcher K.E. and Scivyer C.R. *Quality Assurance (QA) In the UK Building Industry: Its Current Status and Future Possibilities (1988)* CIB W/65 Organisation & Management of Construction Proceedings Volume 1, pp. 367–374, Spon

(9) Hill A.S.B. *J.L.O. Annual Conference (1985)* Source: Atkinson G. 'A guide through Construction Quality Standards' (1987), Van Nostrand Reinhold (UK)

3 Quality Assurance Standards

3.1 Philosophy and Development

The general principles of quality assurance were originally developed in the manufacturing sector of industry to assure the consistently good output of products. Within the UK, the principles of quality assurance have been developed and encapsulated in BS 5750 which forms the document now used as the basis for the application of quality assurance to many different industries including construction. There have been some problems with the performance of buildings and there is a great deal of scope for improving quality in building. Government pressure to improve the quality of UK manufactured products has led to the widespread promotion of quality assurance and part of that pressure is exerted in building through major Governmental departments commissioning construction works. Many aspects of UK construction industry have responded to the requirement for better quality and quality assurance has, therefore, developed beyond the field of materials, components and products and is addressing building design, site assembly and the management of construction.

Formal quality assurance schemes for manufactured goods are well established, well recognised and have been used for many years, although only a limited number of such products are used in the building process. Success in improving quality is dependent upon how well the quality system structure provided is an adequate practical definition of quality. Such schemes are represented well by the BS Kitemark scheme, first used in 1926, and the Agrement scheme, set up in 1966, to cover materials and products where no appropriate BS has been issued.

Although it is desirable to encourage formalised quality assurance schemes for materials and products, based upon clearly defined and recognised standards, it is also recognised that such actions have limited effect upon improving quality within the field of construction.

It is well recognised that only a relatively small proportion of performance problems that are experienced in construction result directly from in-adequate quality in the manufacture of building materials and components, but are influenced more by deficiencies in design formulation, project communication, workmanship, and site management.

BS 5750, published in 1979, provides the basis for introducing quality assurance within these areas. This standard presents principles of management to assure that a firm is well structured and organised to produce goods or services to a required specification. If the firm can demonstrate that it has a quality system that works in accordance with those principles of BS 5750 it can seek recognition and registration as a quality assured company. BS 5750: Quality Systems is addressed in detail in Chapter 5.

BS 5750 has developed separately from the BS Kitemark scheme and denotes quality standards in those areas outside manufactured products. Development of the standard has resulted from continued pressure for improved quality in all sectors of industry and as such, BS 5750 is a generic scheme for all aspects of industry and commerce.

In recent years, a number of firms and organisations have begun to consider the requirements of BS 5750 and application of quality assurance systems to their own activities. Specialist quality assurance schemes are being developed within general construction, private sector housebuilding, building design and specialist sub-contracting. For such operations, the British Standards Institution have introduced their scheme for Registered Firms of Assessed (quality assurance) Capability.

Housebuilding sees one of the UK's most well developed quality assurance systems in the National House Building Council (NHBC) ten-year warranty scheme. Private quality assurance organisations provide quality assurance approval and registration (certification) schemes, such as Yarsley Quality Assured Firms Ltd (YQAF) and Lloyds Register (QA) Ltd. These bodies provide independent assessment and control of quality assurance standards in industry.

The British Standards Institution (BSI) is the largest and most important certification organisation currently involved in the development of quality assurance systems, having led the way towards improving quality standards throughout the 1960s and 1970s, publishing BS 4891: A Guide to Quality Assurance in 1972, A National Strategy for Quality in 1978 and BS 5750: Quality Systems in 1979. BS 5750 is now the UK's National Standard for Quality Systems and is recognised as the basis and general framework for all new quality assurance certification schemes at the present time.

3.2 The Structure for Quality Assurance in the UK

Accreditation and Certification

Any company wishing to become fully recognised for quality assurance must become registered with a 'certification body' and these certifiers in turn must be approved or 'accredited' by Government.

A 'Certification Body' is defined as:

an impartial body, governmental or non-governmental, possessing the necessary competence and reliability to operate a certification scheme and in which the interests of all parties concerned with the functioning of the system are represented.

The term 'accreditation' is defined as:

The formal recognition by a national Government against published criteria, of the technical competence and impartiality of a certification body or testing laboratory.

Certification, 'the act of licensing by a document formally attesting the fulfilment of condition' (BS 4778), is deemed to cover: product conformity certification; product approval; certification of suppliers; management systems; and certification of personnel involved in quality assurance activities.

The National Accreditation Council for Certification Bodies (NACCB)

All certification bodies are controlled by national Government through the Department of Trade and Industry (DTI) and its representative office, the National Accreditation Council for Certification Bodies (NACCB). In essence, the NACCB has responsibility for the assessment and approval (accrediting) of the certifiers (the certification bodies) and provides a uniform system of approval for all UK Certification schemes.

The NACCB is the sole authority for undertaking the assessment of certification bodies applying for accreditation by the DTI. NACCB performs a detailed assessment of the capabilities of the certification body in the same way as the certification body undertakes for the organisation applying to the certification body for a certificate of assessed quality assurance ability.. The NACCB holds a list of accredited certification bodies which includes the particular category of certification and the scope within each category for which the certification body has been accredited.

NACCB issue guidance documents for organisations seeking registration. These documents detail the following:

- Prospectus
- Rules of Procedure for Application
- Criteria for the Assessment of Competence
- Regulations Governing Accreditation
- Guidelines for structure and content of control procedures for certification bodies.

Certification bodies operating certification schemes may carry out assessment of Quality Assurance systems in specific areas. The 'prospectus' lists four categories of certification for which the NACCB are responsible:

(i) Quality Assurance Systems
(ii) Product Conformity
(iii) Product and Services Approval
(iv) Personnel Assessment of Persons Involved with Quality Systems

The prospectus describes how the NACCB processes applications for accreditation, details its assessment procedures, both initial and on-going. It also specifies the costs incurred in accreditation against the initial cost and in monitoring the system in the long term.

'Rules of procedure' specify the Council's terms of reference, its membership body and procedures. The 'criteria for competence' details the basic criteria with which all organisations applying for accreditation must comply. Guidelines on compilation of control manuals is a fundamental document as it describes the requirements with which a certification body must comply to achieve full approval and registration with the NACCB.

As with earlier endeavours in quality assurance, the British Standards Institution played a significant role in the development of the NACCB. The Council was originally created under the BSI's Royal Charter, although it must be emphasised that the BSI acts only as a certification body and not as an accreditation body. Accreditation can only be given by the Government through the NACCB, which means that the BSI, like all certification bodies, is accountable to the NACCB. The NACCB has accredited a number of bodies who provide certification schemes to industry. Application for assessment, registration and full certification is made to these bodies. Three bodies are of particular significance to construction, these being:

(i) The British Standards Institution, Quality Assurance Services Scheme – (BSI/QAS).
(ii) Lloyds Register of Quality Assurance Ltd – (LRQA, Ltd)
(iii) Yarsley Quality Assured Firms Ltd – (YQAF, Ltd).

These bodies have through the broadening of their activity evolved from standards and testing institutions to assume a key role currently. Any group in any sector of industry however may combine to form new certification bodies if the need exists and with reference to NACCB.

The purpose of accreditation by Government is:

(i) To provide assurance to the customer that the sources from which they procure goods or services are themselves a quality source.
(ii) To illustrate to world markets that the UK Government takes a hand in ensuring quality in its national products, goods and services.
(iii) To provide a unified system of appraising QA certification bodies and standards.
(iv) To police certification bodies to ensure quality of their services.

3.3 Sources of Quality Assurance Standards

In the UK, the majority of quality assurance approval, or 'certification', schemes are based upon a structure devised by the British Standards Institution. The BSI's 'Kitemark' certification scheme applies to most products for which there is a British Standard. Where no standard exists, the BSI Register of Firms of Assured Capability provides a measure of guarantee.

The British Board of Agrement (BBA) carries out similar assessment to the BSI and issues certificates for products and processes where no British Standard exists or where products are at an innovative stage in their development. Such schemes are operated on the basis of initial assessment followed by regular policing by inspection visits.

The benefits of certification schemes to the customer are as follows:

(i) The independent and continual assurance that products or services have defined standards and are produced under rigorous quality control.
(ii) To remove the need for the customer to carry out acceptance testing of the purchased goods or services.
(iii) The customer is safe in the knowledge that the product meets a recognised nationwide standard and that quality will not differ between different suppliers.

In the field of materials and products, one positive application of quality assurance, within UK construction, is the scheme operated by the National Measurement Accreditation Service (NAMAS) formed from the amalgamation of the British Standards Institution and the National Testing Laboratory Accreditation Service (NATLAS). NAMAS is a voluntary independent (third-party) accreditation scheme run by Government for the benefit of all types of testing laboratories. Laboratories are assessed, against a set standard, in terms of their competence to carry out specific tasks. NATLAS currently has approximately 500 laboratories accredited within their scheme.

Current scope of accreditation under the NATLAS scheme covers a wide range of construction materials, goods and products. The scheme is based upon documented testing procedures coupled with annual inspections to check that the quality assurance system is constantly in operation and working efficiently. The scheme is very flexible and is frequently adapted to cover the testing of new materials and products where required.

The British Standards Institution founded in 1901 and consolidated by Royal Charter in 1929 is the national standards body for the UK. Under its Charter its function is described as being:

To co-ordinate the efforts of producers and users for the improvement, standardisation and simplification of engineering and industrial materials, so as to simplify production and distribution, and to eliminate the national waste of time and material involved in the production of an unnecessary variety of patterns and prices of articles for one and the same purpose ... and to set up standards of quality and dimension, and prepare and promote the general adoption of British Standard Specifications and Schedules in connection therewith.

Like NATLAS, the BSI, as an important material and product testing house, is responsible for a wide range of testing services to predetermined and carefully controlled standards. Promoting the concepts of quality assurance, the BSI has pioneered the Kitemark system for ensuring the quality and safety of products and has developed the BSI Quality Assurance Services (BSI/QAS) scheme under which the Registered Firms of Assessed Capability operates. It also develops Codes of Practice (CPs), incorporating these into the British Standards upon revision, and is actively involved with European and International standard making bodies. It is from these initiatives that the BSI has championed the requirement for a nationwide standard for quality, (BS 5750), not only in materials and components but also in services to production including design and assembly. It is in these areas that quality assurance can be applied to building design, site construction and construction management. The BSI acts as a certification body for BS 5750: Quality Systems in Construction.

3.4 BS 5750 Quality Systems: UK Standards for Quality

Quality Assurance Standards

Any organisation seeking recognition and formal certification for quality assurance must develop and implement a quality assurance system to the guidelines of British Standard 5750, the UK national standard for quality systems. The standard guides users and those responsible for assessing a firm's quality system (the certification bodies) determine what is required of a quality assurance system in principles and practice. BS 5750 is a standard throughout industry and provides guidance on the control of quality throughout design, procurement, production and commissioning.

BS 5750: Quality Systems, is presented in a number of parts:

Part 0:   Section 01:Guide to selection and use
           Section 02:Guide to quality management and quality system elements
Part 1:   Specification for design/development, procedures, installation and servicing
Part 2:   Specification for production and installation

Part 3:   Specification for final inspection and test
Part 4:   Guide to use of BS 5750: Part 1; Specification for design/ development, procedures, installation and servicing
Part 5:   Guide to use of BS 5750: Part 2; Specification for production and installation
Part 6:   Guide to the use of BS 5750: Part 3; Specification for final inspection and test (1981).

To develop and implement a quality assurance system to meet product or service assurance, any system must meet the detailed requirements of one or more parts of BS 5750, Parts 1, 2 and 3. The standard does not specify a mandatory system but is advisory in giving guidance to users on what should be included in a recognised quality assurance system.

To illustrate the level of detail in the requirements, BS 5750 Quality System Part 1: Manufacture and Installation (1987 edition), for example, presents its requirements in the following areas:

0.   Introduction
1.   Scope
2.   References
3.   Definitions
4.   Quality
4.1   Management Responsibility
4.2   Quality System
4.3   Contract Review
4.4   Design Control
4.5   Document Control
4.6   Purchasing
4.7   Purchaser Supplied Product
4.8   Product Identification and Traceability
4.9   Process Control
4.10 Inspection and Testing
4.11 Inspection, Measuring and Test
4.12 Inspection and Test Status
4.13 Control of Nonconforming Product
4.14 Corrective Action
4.15 Handling, Storage, Packaging and Delivery
4.16 Quality Records
4.17 Internal Quality Audits
4.18 Training
4.19 Servicing
4.20 Statistical Techniques

Each area of requirement is defined by BS 5750 or BS 4778 and a detailed description of instructions given.

For example, Requirement 4.2 states:

4.2 Quality System. The supplier shall establish, and maintain a documented quality system as a means of ensuring that product conforms to specified requirements

Part 2: Specification for products and installation, and Part 3: Specification for final inspection and test follow a similar format and detail of description.

The basic Standard, document BS 5750: Quality Systems (1979) is updated as required with BSI quality system 'Supplements'. For example, the Supplement – Quality Assurance of Products and Assessment of Firms' Capability (1986) details the relationship of Standards to quality assurance practice and outlines the UK quality assurance certification process. It can be seen from BS 5750, Part I that the terminology and structure can create confusion when applied to the building process.

Requirements of BS 5750 Quality System

The adoption of Quality Assurance implies the application of a Quality System – 'the organisation structure, responsibilities, activities, resources and events that together provide organised procedures and methods of implementation', (BS 4778). The development of a quality system must meet the required standards of BS 5750 and it must be clearly demonstrated that this has been achieved through certification. The essential requirement in the development of the system is that it becomes planned, monitored, controlled and documented, in simple terms, plan, implement and record.

Quality Assurance as a practical function becomes more meaningful when the elements needed to address the requirements are appreciated. Key elements are listed below and considered in greater detail in Chapter 5.

- *Quality Systems* development of a policy statement and quality manual and plans of operating procedures and system review by quality assurance management.
- *Organisation* definition of responsibilities of specified functions, including procedures programme of implementation and working plans, under the direction of the quality manager.
- *Auditing* periodic review of effectiveness of the quality system by independent auditing from within or without the organisation.
- *Planning* detail of specific operations and setting of standards and specifications which define quality required.
- *Instructions* preparation and use of standardised procedures and documentation format.
- *Records* written account of system implementation and feedback required.
- *Controls* formal method of identifying problems and taking

corrective action to eliminate non-conformity to quality requirements and to record any changes to the quality system. Control can be sub-divided into design, procurement and equipment, as follows: (1) Design: (use of recognised codes and design guides); (2) Procure-ment: (use of accredited materials and services); (3) Equip-ment: (accuracy checks on inspection methods and test procedures).

- *Production* final inspection: use of sampling and inspection to assure work in progress and quality of final product or service.
- *Protection* use of standard method for handling, storage and protection to preserve product quality.
- *Training* system of training of personnel in all functions affecting quality.

System Terminology

To fully appreciate the elements of a BS 5750 Quality System, it is useful to consider the terminology of specific aspects. The terms used can be defined as follows:

- *Quality Policy* The overall intentions and direction of an organ-isation regarding quality as formally expressed by company manage-ment.
- *Quality Manual* The Quality Manual is a 'document or set of documents, setting out the general quality policies, procedures and practices of an organisation'. (BS 4778). The manual should specify the structure of the organisation and responsibilities of personnel involved in maintaining the quality system.
- *Quality Management* That aspect of the overall management function that determines and implements the quality policy.
- *Quality Procedure* The Quality Procedures of an organisation are the documents which describe the activities involved in pursuing the achievement of quality. Quality Procedures should be method state-ments which make reference to relevant specification documentation.
- *Quality Programme* Defined by BS 4778 as 'A documented set of activities, resources and events serving to implement the quality system of an organisation'. A Quality Programme is a written description of a Quality System.
- *Quality Schedule* This is a document which specifies the essential items of BS 5750 which must be included in a Quality Plan for a specific product or process, and, therefore, its scope is limited.
- *Quality Plan* This is 'a document derived from the Quality Programme setting out the specific quality practices, resources, and activities relevant to a particular contract or project'. (BS 5750).
- *Quality Audit* 'The independent examination of quality to provide information. Quality auditing can relate to the quality of a product,

process or system. Quality auditing is usually carried out on a periodic basis and involves the independent and systematic examination of actions that influence quality. The object is to ascertain compliance with the implementation of the quality system, programme plan, specification or contract requirements and where necessary their suitability'. (BS 4778).

- *Quality Documentation* Quality Documentation is defined within this Quality Plan. Documentation includes all records and documented evidence necessary to demonstrate adherence to the Quality Plan. The Quality Plan and record of implementation make up written records of the quality system in operation.
- *Quality Surveillance*   This is defined as the systematic monitoring and verification of the use of procedures, methods, products and services and the analysis of records in relation to set standards to ensure that the requirements for quality is being complied with.
- *Quality Control*   Control is defined by BS 4778 as 'The operational techniques and activities that sustain the product or service quality to specified requirements. It is also the use of such techniques and activities'. Quality Control (QC) is an inherent aspect of quality assurance involving the inspection, testing and analysis of the quality of the process being undertaken.
- *Quality Manager* The Quality Manager is a designated senior manager who assumes responsibility for ensuring that the organisation's Quality Policy is documented, implemented and controlled.
- *Quality Review*   This is the formal system assessment and modification where necessary, by senior management of the status of the Quality System in implementation in relation to their set Quality Policy.

3.5 Quality Assurance: Types of Assessment

The principles of quality assurance can be demonstrated in three different types of quality system, 'Third Party' or full independent certification being the most detailed. This type, is therefore, the highest level of quality system implementation.

Many organisations do not require such detailed quality systems or full certification to secure the benefits of a quality management system and therefore, implement only the intermediate stages of a full certification programme. The level of assessment to which an organisation aspires is, in real terms, influenced by its need and commitment, in fact it remains an organisational prerogative. It is important however, that the organisation recognises the need for some form of assessment whether it be first, second or third party in nature. Every contribution to quality assurance can only help improve quality within the building process.

There are essentially three types of quality assurance systems. These are as follows:

(i) *First Party Assessment* This involves setting up a quality system, documenting the structure and procedures and subsequently notifying the client that a quality system has been implemented.

(ii) *Second Party Assessment* This involves developing a quality system and inviting the client to collaborate in its development to fulfil the requirements for quality as specified by the client. Alternatively a client may carry out its own assessment of an employee's QA system.

(iii) *Third Party Assessment* This is independent 'full' certification (assessment and registration) under the guidance of BS 5750: Quality Systems. This form eliminates multiple assessments.

Any organisation seeking full or third-party certification must develop their quality assurance system beyond the first two levels of detail, or first-party and the second-party level and become registered with an independent certification scheme accredited with the Government through NACCB. Only such quality assurance schemes are a conclusive guarantee of an organisation's commitment to quality.

4 Quality Assurance: Certification

4.1 The Method of 'Third Party' (Full) Certification

Third-party quality assurance certification can be approached in two broad ways:

(1) *Certification Under a Registered Industry Sector Scheme* Some industries have authority under the NACCB to undertake certification where an industry sector group exists demonstrating recognised mutual needs.

(2) *Direct Certification* Where an organisation, not devoted to a single field of activity, approaches an independent body directly for third-party assessment.

The majority of organisations within the field of building and construction will apply for direct certification to an independent body through assessment, approval and registration of their scheme.

Certification Bodies

Some of the major certification bodies, accredited by the NACCB, include the following:

- British Standards Institution: (BSI)
- Lloyds Register Quality Assurance: (LRQA)
- Yarsley Quality Assured Firms Ltd: (YQAF)
- British Approval Service for Electric Cables: (BASEC)
- Ceramic Industry Certification Scheme Ltd: (CICS)
- UK Certification Authority for Reinforcing Steel: (CARES)
- Timber Research & Development Association: (TRADA)
- Water Industry Certification Scheme: (WICS)

Of these bodies, three are of particular relevance to the construction industry, these being: BSI, Lloyds Register for Quality Assurance and Yarsley Quality Assured Firms Ltd.

British Standards Institution

The BSI has two broad areas of work within its remit:

(i) Certification and Assessment
(ii) Inspectorate

Under 'Certification and Assessment' the BSI is responsible for management of the BSI Kitemark (product certification) Scheme and Safety Mark Licensing (product conformation to British Standards specifically concerned with safety). The second area, 'The Inspectorate' provides a range of assessment and surveillance of products and quality assurance systems and maintains the BSI 'Register of Firms Assessed Capability'.

The BSI/QAS is a commercial organisation offering certification services to industry in competition with other certification bodies. It is incorporated within BSI's framework for administration purposes only, having origins in BSI historically.

The BSI Kitemark Scheme

It is important to stress that the BSI scheme to register firms of assessed capability or quality to BS 5750: Quality Systems is not the same as the BSI's Kitemark Scheme. These two aspects are sometimes confused. Kitemarking is product specification. The Kitemark is a registered 'trade mark' owned by BSI. It can only be used by product manufacturers licensed under a particular BSI Kitemark Scheme. The Kitemark indicates the BSI has undertaken a test of samples of the product under consideration and assessed them against the requirements of the appropriate British Standard and confirmed that the standard has been fully met.

The Kitemark can, therefore, not be applied to an organisation, or to a structure built by a quality assured and registered firm. Confusion arises since a contractor, registered under the BSI/QAS scheme is permitted to use the BSI's Registered Firms logo which is similar to Kitemarking.

The BSI Safety Mark Scheme

The BSI Safety Mark Scheme is a similar scheme to the Kitemark process, but is limited to manufactured products which conform to British Standards specifically concerned with safety, or to the safety requirements in other standards. Products meeting the assessment of BSI are entitled to use the designated BSI Safety Mark.

BSI/QAS Scheme

Product manufacturers are also required to develop and maintain a quality system based upon BS 5750: Quality Systems which is also assessed by BSI as an integral part of the product certification process. This specifies the organisation of responsibilities, procedures and methods involved in manufacturing the product and is the same as the quality system developed by organisations providing services to industry under BS 5750. The quality system scheme – BSI Registered Firms System has become an increasingly growing and important area of BSI's quality assurance and certification work.

To simplify the role of BSI in the certification process, an organisation can apply to the BSI for:

(i) Product specification under the Kitemark Scheme where products meet the British Standards.
(ii) Product specification under the Safety Mark Scheme where products meet the British Standards specific to safety.
(iii) Quality Assurance Services, (BSI/QAS), represents quality assessment under the BSI Registered Firms Scheme where quality assurance systems for product manufacture or services (design, construction, consultants, and so on) meets with the requirements of BS 5750: Quality Systems.

Certification Approach

The BSI/QAS scheme is a useful example to describe the general approach to certification. There are key points in the BSI/QAS system which, by assessment and continued surveillance, provides an independent assurance of an organisation's capability of working to specified requirements. The system documents to assess all firms in manufacturing processes or service industries, including construction, must follow BS 5750: Quality Systems. If an organisation's design capability is to be included in the assessment, BS 5750 Part 1 applies. An organisation that works to a published specification or to the client's or customer's specific requirement is assessed under Part 2. The exact scope of assessment is determined by the organisation's range of services. In addition to BS 5750 Parts 1, 2 and 3, a BSI Quality Assessment Schedule, (a document outlining the essential items from BS 5750 to be included in an organisations Quality Plan for a specific product, process or service), must be complied with.

There are four major stages to certification under the BSI/QAS scheme:

(i) *Formal Application* organisation submits written application and documents to BSI.

(ii) *Initial Assessment*   quality assurance system documents are assessed against BS 5750.

(iii) *Issue of Certificate*   if the quality system meets the requirements of BS 5750 a certificate of registration is issued.

(iv) *Surveillance*   periodic checks are made to continually assess the quality system in operation.

An organisation seeking registration must have a fully worked and documented quality system which complies with the appropriate part of BS 5750 and the related Quality Assessment Schedule. A written application is submitted to the BSI, together with documentation detailing the quality system to be assessed. The route to quality assurance certification is shown in Figure 4.1.

Types of Registration

Prior to assessment the BSI compiles a detailed appraisal of the applicant's documentation. The documented quality assurance system is assessed against the appropriate part of BS 5750 and in the case of products certification, the Quality Assessment Schedule for that range of production. Once the documented procedures are passed as satisfactory, an assessment visit is arranged to the applicant's organisation to assess the quality system and control procedures on site. This is to ensure that their practical implementation meets with the principal standards required. At the conclusion of the assessment, the assessor makes a verbal recommendation of initial registration or non-registration at a final meeting with the organisation's quality assurance representatives. This is supported by a written summary presented for discussion, during which any difficulties and discrepancies can be considered.

There are three possible recommendations:

(i) Unqualified Initial Registration – where no discrepancies in the system have been identified.

(ii) Qualified Initial Registration – where minor anomalies have been highlighted which can be easily rectified. 'Minor' problems are considered as being neither individually nor cumulatively serious in nature.

(iii) Non-Registration – where there is a lack of system or procedure or when major alteration to the system is demanded.

Following assessment, if all quality procedures are found to be satisfactory, a written confirmation is sent to the organisation. Where 'unqualified' or 'qualified' initial registration is awarded, a certificate of registration is issued to which is appended the scope of the organisation's initial registration. The registration assessment process continues, subsequent to

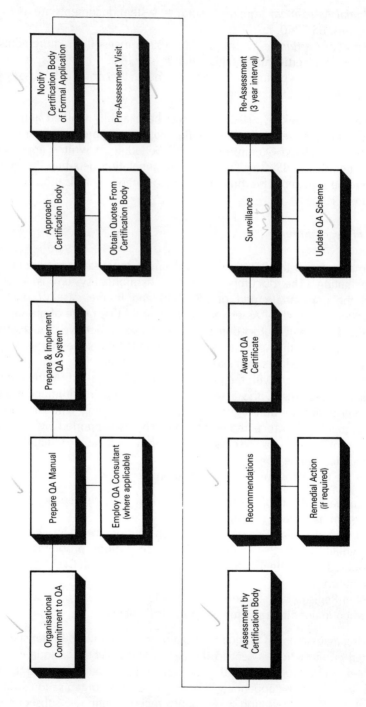

Figure 4.1 Route to quality assurance (third-party) certification

the award of an initial registration, to confirm that the quality system is operating efficiently and effectively. Where qualified initial registration has been awarded, the organisation is required to rectify discrepancies and problems with their system within a stipulated period. Registration can then be upgraded to unqualified status.

At the end of the initial registration assessment process, and once the assessor is satisfied that the quality system is operating effectively, the registration is confirmed. The organisation is then entitled to display the BSI's Registered Firm Symbol on their company documentation and literature. Organisations are monitored periodically after registration through routine, but unannounced surveillance visits, usually a minimum of two per year, to ensure that the quality control standards observed during the initial assessment phase are maintained.

The BSI/QAS scheme, with the Register of Firms of Assessed Capability is recorded in the BSI's 'Buyers Guide' and are continually updated in the BSI publication 'BSI News'. There are, in addition, many organisations involved in building and construction who have already implemented, or are preparing, schemes for certification. Whilst these schemes must follow the guidelines of BS 5750: Quality Systems, not all are certificated by BSI/QAS but represent those alternative schemes provided by Lloyds Register Quality Assurance and Yarsley Quality Assured Firms Ltd.

Independent Certification Bodies

Yarsley Quality Assured Firms Ltd

Yarsley Quality Assured Firms Ltd (YQAF) is an independent quality assurance consultancy and certification organisation, accredited by the NACCB.

Yarsley provides two main services to industry in connection with quality assurance:

  (i) Quality Management Consultancy – offered through 'Fulmer Yarsley', formally the Yarsley Technical Centre.
 (ii) Quality Management System Certification – 'YQAF Ltd'.

YQAF Quality Management System Certification is a quality assurance scheme designed to provide independent assessment and certification of an organisation's quality system to the appropriate parts of BS 5750: Quality Systems, and the relevant Quality Assessment Schedule (QAS). The procedure of seeking registration is similar to that detailed for the BSI/QAS scheme. After initial assessment, a certificate of registration is issued for a three-year period. Surveillance visits are made twice yearly over the three-year certification period to ensure continual compliance to BS 5750 and the Quality Assessment Schedule.

Yarsley is committed to practical implementation of BS 5750 and its European and international equivalent standards. It operates in accordance with the NACCB Criteria of Competency with its standard of certification following BS 5750 and International Standards Organisation (ISO) 9000 Series of Standards.

Registered Firms are issued with a certificate of registration and are entitled to use the YQAF logo on organisational documentation and promotional literature within the scope of cover specified in an Assessment Schedule bearing the certificate number.

YQAF has around 30 companies currently registered within the scheme in a wide range of material supply, product manufacture and engineering fields. Yarsley also provides extensive services to industry through its 'Quality Management Consultancy'. YQAF consultants provide guidance and assistance in developing and documenting quality systems. Working closely with trade associations and industrial training boards, YQAF Ltd, provides consultants fully versed in the principles of quality assurance to BS 5750/ISO 9000 and in applying these principles to practice. Quality Assurance services can involve; pre-certification examination of an applicant's existing quality scheme and control systems against the applicable parts of the standard; developing action plans to produce quality manuals; and assistance with preparation for a full certification application. Training services are also available including an introduction to quality management, quality systems education, internal audit procedures and control management techniques.

YQAF Ltd also provide 'Product Conformity Certification' under their 'Testguard' scheme. Testguard is an independent evaluation of a product and establishes the product's ability to meet certain operating criteria and standards such as: British Standards; German Standards Institute (DIN); ISO; a company's own specification; and individual purchasers' specifications.

Following application and initial assessment, samples are tested within Yarsley Laboratories or an alternative recognised testing facility such as the National Measurement Accreditation Service (NAMAS). If the tests are satisfactory, a certificate of registration for the product is issued. Periodic surveillance visits follow as for the certification of services, previously described. The product may use the Testguard logo on its packaging and promotional literature to the guidelines of use issued by Yarsley.

Lloyds Register of Quality Assurance

Lloyds Register of Quality Assurance Ltd (LRQA), like Yarsley Quality Assured Firms Ltd, is an independent third-party certification body accredited by the NACCB. LRQA provides a number of certification schemes for product based industry and single firm certification schemes for quality management systems. Lloyd's quality management system certification follows the principles used in the Yarsley and British Standards Institution schemes with initial assessment leading to registration followed by subsequent surveillance and site visits. Like other schemes, the LRQA follows the guidelines of BS 5750 and ISO 9000 series of standards.

LRQA Ltd, have produced a 'Guidance Document to Quality Systems' outlining to prospective applicants for certification what is expected of the quality system and standards. This document is supplemented by a series of 'Quality System Supplements' relevant to particular sectors of industry. In the field of construction, for example, there are Quality System Supplements addressing: Management in the Construction Industry: Project Management; Engineering; Construction and Installation Organisations; and the like. These set out the specific requirements of the quality system in their respective categories and are read in conjunction with the standard Guidance Document. Lloyds also provide pre-assessment quality assurance consultancy services developing an application from an outline quality policy and compilation of manuals through to assessment, certification and surveillance.

4.2 The Value of Quality Assurance Certification

With an effective quality management system, any company should experience improved organisation and operational efficiency. Organisational problems may be reduced and faulty product manufacture can be prevented, not merely inspected and rejected. Improvements can lead to considerable financial savings for companies in the medium to long term.

Certification of an organisation's quality system will bring national and international recognition of the organisation's commitment to quality. Its marketing benefits are clear as it shows the customer or client that the organisation has received an independent assessment of its management and production procedures and that the output service or product is of an unequivocally high standard of quality. Quality systems also have the benefit of 'getting the job done right first time'. It can prevent shoddy work; reduces organisational wastage; and the formalised quality assurance procedures can assist in the smooth running of the organisation.

Direct advantages of certification arise from:

- Registration in a national and international system of standards for quality.
- Use of a recognised registration logo.
- Listing in the Department of Trade and Industry (DTI) Register of

Quality Assurance companies and the listings of equivalent bodies, not only in Europe but worldwide.

- Registration in a product conformity assessment list (such as Yarsley's Testguard, where applicable).
- Registration in a list of companies made publically available by BSI, Yarsley and Lloyds respectively.
- Long-term assurance to the client/customer that quality of product or service will be maintained to a standard by continued surveillance, control and up-date of the system.

4.3 Cost and Time Implications of Certification

Costs

Potential benefits, from establishing and maintaining a certificated quality assurance system, are not secured without cost to the organisation. There are direct costs resulting from the certification process itself and the indirect, or somewhat hidden costs of structuring the quality system within the organisation.

Significant costs are incurred in:

(i) Developing the quality assurance system.
(ii) Producing the quality policy, manual and plans.
(iii) Setting up the system implementation.
(iv) Maintaining an internal audit system.
(v) Independent third-party assessment and certification.

Cost of quality assurance assessment and registration varies between certification bodies although the BSI, Yarsley and Lloyds all publish their fees for certification services. Direct fees from a certification body includes a proportion to cover the cost of NACCB accreditation and maintenance and again these fees vary amongst different certification bodies.

More difficult to assess are the indirect costs to the organisation seeking certification. There will be costs for: developing and maintaining a quality system; employing the necessary quality assurance staff or consultants and providing training; liaising with the certification body; the preparation of documentation; and for operational changes to encompass the quality system as certificated. Further hidden costs within the organisation are likely to arise from changes in maintaining records, increased levels of supervision, and the like. Following successful assessment and certification, costs continue to be borne in the fees of the certification body maintaining their surveillance visits and checking procedures.

Costs can be reduced in certain circumstances, for example, if there is a possibility of a group of firms collaborating in the design of documentation, although each firm will ultimately have to become assessed and registered separately. Some organisations can also apply, under certain conditions, to the DTI for financial assistance with quality related initiatives. Some quality assurance schemes, such as Yarsley, also offer the remittance of direct cost over a number of years to ease the initial cost burden to the organisation.

An approximation of total cost within building, design or other construction professions, is somewhat difficult to generalise since each organisation must develop its own unique quality system, given the particular nature, scope and remit of its operations and organisation. Within construction, quality assurance certification may ultimately become a mandatory requirement. Although voluntary at present, organisations servicing larger public and commercial clients must provide a certificate of competence as a condition of contract, and as such, must treat organisational costs incurred in establishing their quality system as a necessary overhead. Most organisations will offset the cost implication of quality assurance by the improved marketing benefits derived.

Direct certification fees are easier to assess as certification bodies specify their various registration fees. Fees are subject to some variation depending upon the size of firm and the number of its activities that need to be registered. Building companies, for example, pay a basic 'Application Fee' under the BSI scheme. They also pay a separate 'Assessment Fee' based upon the time required by the assessment team to evaluate the company and must also pay an 'Annual Certificate Fee' and subsequent 'Surveillance Fees'.

Table 4.1 shows typical certification fees (costed as at January 1990). The figures should only be interpreted 'as a guide' as variation exists amongst various certification bodies. Fees may vary due to some of the following factors:

- Size of company and number of employees
- Structure of the organisation
- Diversity and range of the company's activities
- Nature and complexity of the quality system
- Complexity of documentation.

Time

The timescale to develop, implement, audit and refine a quality system in order to be able to apply for a quality assurance certification, will vary depending upon the size of the organisation, the operating procedure and its workload. Quality systems developed from scratch through to receiving a certificate of registration may range from one year to eighteen months for a relatively small organisation.

Table 4.1 Estimate of typical quality assurance certification fees

Fees per annum	£		
Application fee	*500		
Assessment fee	1 200		
Surveillance fee	1 000		
Option 1			
Year	1	2	3
Application fee	500	–	–
Assessment fee	1 200	–	–
Surveillance fee	1 000	1 000	500**
	2 700	1 000	500
Option 2 (spread cost)			
Year	1	2	3
Application fee	500	–	–
Assessment/			
Surveillance fee	1 300	1 300	1 100
	1 800	1 300	1 100

*Reduction for companies employing fewer than, say, 10 persons.
** In year 3, the surveillance visit may be replaced by a re-assessment appraisal to renew certification for further three-year period. All expenses may be charged extra, at cost, and all fees and charges may be subject to VAT at standard rate. Additional charges are also made for work undertaken arising from identification of non-conformances and unscheduled follow-up work.

The first three to six months may be spent in identifying and developing the basic procedures and documentation required by the certification body, in accordance with set criteria. The next three to six months is likely to be used to assess a trial run of the procedure, identifying and rectifying discrepancies. A further three months will be spent liaising with the certification body, and a final month or more undergoing initial and final assessment before a certificate is obtained.

To acquire certification, an organisation must be able to demonstrate and provide evidence that the quality system is well developed, is effective and operates throughout the entire organisation from company board level, through senior management to supervisory management at the workplace. During the initial stages of development, participation by the company board and assigned Quality Manager, (responsible for maintaining the quality system and ensuring the process is on-going after certification is obtained), senior management and quality assurance consultant, where employed, is essential in formulating the basic policy, principles and procedures of the system. Whilst it is not possible to determine the actual time involved in the quality assurance process by these various levels of management, it is possible to illustrate their respective contribution to the process. This is shown in Figure 4.2.

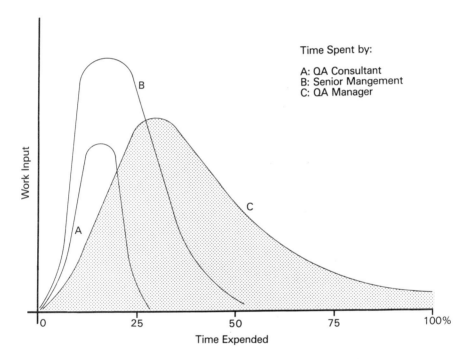

Figure 4.2 Distribution of work input by organisational management in development and maintaining quality assurance system. Adapted from Quality Assurance: Introductory Guidance RICS (1989)

Costs and Benefits

The degree to which quality assurance is desirable in any particular situation can only be determined by specific, and very detailed cost-benefit analysis. Whereas direct costs of certification are easy to determine, the benefits, although easily identified in terms of potential, are difficult to quantify. At the design stage, it is difficult to ensure that real quality in terms of value for money judgements are made since the result is measured considerably later in the building process by how well the building performs and the cost of repairs and maintenance. It is difficult also to quantify the intrinsic quality of a well designed and well constructed building for the same reasons and, moreover, it is virtually impossible to quantify the satisfaction of the client or building user.

It is becoming clear within the construction industry that importance is being attached to the cost of claims arising from contractual problems and defective construction and it is this concern, as much as any other factor, which is encouraging construction industry to address the quality issue.

Certification gives considerable benefit to an organisation where recognition for quality assurance enhances its reputation and standing, and in particular where the organisation needs quality assurance recognition to maintain or improve its position in the marketplace against stiff competition. In such situations, cost becomes somewhat immaterial to necessity.

Given these considerations, it is more appropriate, not to regard quality assurance as a balance between cost and benefit, but to define quality assurance cost as a basic necessity in producing the product or providing the service to the required standard and appreciate the benefits as a 'bonus' which may be realised if the job is done satisfactorily.

Certification and Indemnity Insurance

A perceived longer-term benefit of adopting quality assurance practices and obtaining registration is a reduction in premiums to professional indemnity insurance policies. Premiums are unlikely to be reduced until there is considerable and unambiguous evidence to show that the number of claims made against organisations operating quality assurance systems are less then those made against organisations without quality systems. It is too early in the evolutionary process of quality assurance to expect insurance companies to meet this expectation in the immediate term. It is possible that an independent quality assurance organisation could lose its accreditation for failing to adhere to the system that they have created to meet the specific requirements of a third-party certification body. This could give insuring underwriters considerable cause for concern and therefore raise premiums. This situation is however, a most unlikely example as accreditation could only be lost following a very serious breakdown in the quality assurance system. As more organisations seek registration with the available schemes and quality systems become commonplace in the future, there is reason to expect that certification programmes and insurance schemes will be interactive in nature but this remains a long-term expectation.

Certification and Public Recognition

The construction industry continues, in many respects, to have a poor image with the general public. As construction industry is changing to meet the needs of increased requirements for quality, so the public perception of quality is also changing. The public has become more aware and discerning generally in the choice of goods and services and this extends into the construction market.

In housebuilding for example, the market has become more demanding of quality as the buyer expects better value for money. There is little doubt that such demand-pull is improving quality and giving rise to better building practices in some sectors of construction industry.

Certification also seeks to capitalise on such improvements. Organisations that become registered for quality assurance practice can find that they receive improved public recognition and this support is being applied to assist the customer. One UK housebuilder has already devised the original concept of seeking preferential mortgage rates for buyers of houses built by quality assured builders. This is one early example of how quality assurance can bring major benefit to the public and lead to better public satisfaction.

4.4　Other Organisations Associated with Quality Assurance Certification

The British Board of Agrement (BBA)

The British Board of Agrement (BBA) is involved in the testing, assessment and certification of products used within the building and construction industry. The types of product assessed are wide ranging but are generally characterised by being new or innovatory in nature although existing products may also be assessed.

BBA is an independent company limited by guarantee and controlled by a council appointed by the Secretary of State for the Environment and is based at the Building Research Establishment.

An Agrement Certificate gives an assurance that a product or system to which the Certificate relates, if properly used in accordance with the terms of the Certificate, will meet the relevant requirements. In addition to examining product performance, the BBA also approves installation of mechanical systems, for example, domestic water and heating systems, and the like. Scope of assessment of product is defined in five categories: safety; habitability; durability; practicality and maintenance.

The BBA undertakes to assess products and systems where there is no relevant British Standard, or if there is a British Standard, to examine innovative aspects of the product's use. Testing of products or systems is always undertaken at NATLAS accredited laboratories or test centres. BBA adheres to all the requirements of quality systems as specified in BS 5750 and product manufacturers seeking certification must meet these standards. Following assessment and certification, products are monitored by the usual quality system surveillance checks during the period of validity of the certificate, after which re-assessment is undertaken.

The BBA work closely with the BSI. A joint BSI/BBA Liaison Committee maintains and develops co-operation between the two bodies and the BBA aids the Quality Assurance Division of BSI to develop current appraisal, inspection and control schemes in connection with the BSI/QAS certification scheme.

5 Quality Assurance Systems

5.1 The Structure of Quality Systems

Philosophy

Quality Assurance is concerned with planning and developing the technical and managerial competence to achieve the desired organisational objectives. Quality Assurance is also concerned with the management of people, addressing the roles, duties and responsibilities of individuals within the organisation. Whilst quality assurance is primarily the responsibility of management, its structure and implementation must become a part of the total organisational framework. Moreover, quality assurance must also be an important aspect of the marketing and promotional strategy of the organisation. Only when quality assurance pervades the entire organisation and becomes an integral and recognised aspect of its operations will quality assurance foster the potential to become truly successful.

Policy

Where quality assurance is to be adopted within an organisation, there is a need for a clear statement of overall policy and the clear definition and designation of responsibilities for its operational elements. Quality assurance is, in many ways, a matter of integrity and the desire to provide a quality product or service. If that fundamental interest is absent or ignored, quality assurance will simply not succeed. Within an organisation 'Quality Systems' must be implemented from the top down, from executive management through the complete organisational hierarchy. Often the introduction of quality assurance principles will challenge long-established methods and customs and there may be considerable changes required within the organisation.

Organisational policy must be sufficiently solid to impose change without making the transition to new working methods uncomfortable for the employees. Successful quality assurance requires support from everyone in the organisation and policy must, therefore, promote this.

Framework

Within the organisation, a quality framework must be developed. This framework will form the basis of the organisation's approach and provide instruction and information to those involved. Such a framework will include the following:

- The corporate objectives and philosophy towards quality and its assurance.
- A policy statement from the organisation's board of executive management.
- Documentation including: (1) Standards; (2) Plans; (3) Procedures.
- Leadership, motivation and training to provide information instruction and describe responsibilities and accountabilities.

Objectives

Different industrial and commercial markets require different policies, and organisational approaches in meeting the needs of quality assurance. Companies involved in advanced engineering such as the nuclear power industry, and off-shore engineering have, for many years, been aware of the need and practised formal quality assurance through sophisticated quality systems. General contracting within the construction industry is somewhat different in structure, scope and methods of working. It is, therefore, not sufficient to promote simply generic based systems. Rather, each organisation within its own industry sector and specialised field should develop its own quality system, based upon the generic standard framework of BS 5750 to meet its own objectives and the requirements of the commercial market in which it operates.

Documentation

Most organisations operate a quality assurance system of some kind, through well established and good organisational practices. The main difference between such informal systems and the formal arrangement specified in BS 5750 is that the quality system must be documented. The prime objective in developing a quality assurance system is in creating the necessary documentation. Documentation in a quality system makes it amenable to control by management. Unwritten systems suffer from a lack of information and instruction and without such written guidance there is no yardstick against which to measure and control the system.

A useful feature of BS 5750: Quality Systems is that it presents basic guidelines and advice and the quality system can, therefore, be as simple or as complex as the organisation requires. In this way, each organisation can specify its own unique objectives and develop a quality system to suit. Equally, quality systems can be structured to suit all companies, irrespective of their type and size.

Documenting policy and objectives within quality systems is especially important in an industry such as construction where there is a high turnover of staff. Where employees come and go regularly, there is little time to absorb and implement company policy and procedures and verbal communications simply cannot be relied upon. Moreover, client needs, customer demands, design and construction methods and materials are all constantly changing. Quality assurance management systems require constant update to meet changing circumstances. In-built flexibility in basic quality system documentation, allowing revision to policy statements and updated manuals, when required is a useful mechanism for communicating and controlling such change.

Responsibilities

A quality system must become an integral part of the entire organisation if it is to be successful in application. Management, responsible for the organisation's direction and operation, must provide not only the basic policy and structure but provide leadership, instruction, motivation and resources to implement the quality system.

Responsibilities and accountability for quality assurance must be clearly defined and indicated to both line management and for service functions. This must become part of the company's documented organisational framework. It is likely that one person will be designated responsible for quality assurance within the organisation, reporting directly to senior management. This is the function of the 'Quality Manager'. Responsibility, in practical terms however, is not the sole task of the quality manager as everyone in the organisation should give a duty of care to quality in the course of their work.

Once a quality system has been developed, establishing the system and seeking external independent recognition is vital. Whether recognition comes from a client or customer or independent certification body, there must be a formalised and documented method of communication and a point of liaison with the assessing party established by the organisation. There must also be a system of communication within the organisation bringing the policy and procedure to the attention of staff and mechanisms established for the training of personnel. The quality system must ensure that the procedures are updated correctly and that individuals are sufficiently well informed and motivated to implement them.

Management

BS 5750: Quality Systems advise that, as part of any quality system, the organisation should appoint a 'Quality Assurance Manager' to assume responsibility for the overall strategic policy and administration of the quality system. The quality assurance manager, therefore, acts as an intermediary between executive (board) management and quality control management at the production/service level (see Figure 5.1).

The key role, duties and responsibilities of the Quality Assurance Manager can be summarised as follows:

(i) To be responsible for the quality assurance system throughout the organisation and be accountable to the company's board of directors or other quality assurance executive as appointed.

(ii) To compile an organisational policy and procedure handbook to the company's quality policy and objectives, and implement the quality system.

(iii) To be responsible for the monitoring, control, internal auditing and evaluation of the quality system and also to provide reports to executive management.

(iv) To organise the training of all company personnel in aspects of quality assurance for which they are responsible.

(v) To liaise with the independent certification bodies prior to and during assessment, during certification and during surveillance under the period of registration.

(vi) To compile the company's promotional and publicity material and to liaise with external bodies such as clients, customers and the public.

(vii) To evaluate the capability for quality assurance in all suppliers and sub-contractors to ensure they meet with the standards of BS 5750 and associated quality assessment schedules.

Organisation

The type, operation, size and available resources of an organisation will all determine how the quality system is to be developed and implemented. Size is perhaps the most inhibiting factor for quality systems, yet irrespective of whether an organisation is large or small, systems can be developed successfully given organisational drive and commitment.

If the organisation already has an operating handbook, set of standing instructions or written procedures covering the various functions which affect quality, then the first step towards a quality system will have intrinsically been made. If no procedures exist, then it is the Quality Manager's task to identify which aspects require attention and produce written procedures of how these activities must be carried out and whom within the organisation will be assigned responsibility. This aspect is useful to the organisation since the very task of gathering facts, ideas and thoughts and the evaluation necessary to produce written instructions can be creative in highlighting areas of organisational weakness and inefficiency. Restructuring for 'Quality Management' in this way can bring acceptability to change within the organisation.

Further organisation will involve rationalising activity to create defined and documented procedures against which current activity can be compared. Moreover, activity can be assessed against the quality system standard with which the company wishes to comply. If any discrepancies are identified, decisions can be made as to how they may be handled and corrective action be introduced into the normal running of the organisation.

Commitment from management cannot be over-emphasised. The philosophy and policy must come from the highest level of management and made to permeate the entire organisation through a structure that perceives and promotes quality as a way of business.

Implementation

Every organisation will implement quality assurance practices in a way which suits its own individual, even unique operational structure and characteristics. There are however, a number of principles which influence any organisation when developing its quality assurance policy and procedures, these include:

(i)   Quality involves all organisational processes from initial design, through products or service to the satisfaction of the client.
(ii) Quality assurance is the shared commitment and responsibility of all levels of management and the workforce.
(iii) Leadership, motivation and impetus must come from the top and be made to flow inherently through the organisation.
(iv) Each employee must understand the need for quality and accept their part in its overall achievement.
(v)   Policy, procedures, duties and responsibilities must be clearly stated to all concerned.
(vi) All activities must facilitate measurement and assessment (audit) to ensure the objectives are met.
(vii) The system must be sufficiently rigid to promote formality and rigour but remain sufficiently flexible to accommodate organisational or environmental change.
(viii) Overall responsibility to be vested in one individual to co-ordinate the successful achievement of quality. This individual is the 'Quality Assurance Manager'.

5.2 Guidelines to Quality Assurance Systems Development and Implementation

An organisation wishing to progress along the path to registration and full third party certification must develop a quality assurance management system in keeping with appropriate standards. This must demonstrate unequivocally that sound principles and practices of quality assurance have been employed. Although the British Standards Institution, Yarsley Quality Assured Firms Ltd and Lloyds Register of Quality Assurance Ltd all provide their own individual certification requirements for quality systems, a generic trend for requirements can be identified and forms the basis for developing the system and necessary documentation. BS 5750 includes the definitions and descriptions of such basic requirements. These were considered in Chapter 3. Key aspects of quality systems are as follows:

Quality Manual

All procedures are required to be documented in a 'Quality Manual'. This manual is the document setting out the general quality policies, procedures and practices of an organisation. The manual should also cover the structure of the organisation and responsibilities of personnel.

The manual should be presented in the following format:

- Table of Contents
- Revision List
- Distribution List
- Statement of Authorisation by the company's board, executive or managing director
- Summary of the Manual and Instructions on use
- Policy Description
- Organisational Structure
- Auditing Procedure
- Index

Within the Quality Manual, the following elements should be clearly described:

Revision List
To facilitate periodic update a table of revisions and amendments should be provided and include:

- section of the manual/system update
- page numbers of manual concerned
- date of revision.

Distribution List
Each copy of the manual should be numbered and a register kept of manual holders. This will facilitate circulation of updates. Outdated copies should be returned to the quality assurance manager or destroyed.

Statement of Authorisation
The manual should contain an official statement by the company's board, its chairman, chief executive or quality assurance director. This will indicate the overall company philosophy and commitment to quality.

Summary of Manual
Details addressing format, scope and use of the manual should be clearly specified.

Policy
There must be a clearly defined policy relating to quality assurance within the organisation. The policy should include a statement on organisational objectives, responsibilities and commitment to the quality system. It is important that the policy expressed in the quality manual is advised to all personnel within the company so that individual responsibilities are clearly understood at all organisational levels.

Organisation
The organisational framework should be described, and illustrated where necessary, highlighting the structure of company's sub-groups, where so divided. Particular quality management frameworks should be described and the organisation should nominate a quality manager with suitable qualifications and experience to manage the system. This person is responsible for all aspects of quality assurance implementation and is accountable to senior management.

Programme and Procedures
A detailed 'Quality Programme' and list of 'Quality Procedures' should be provided and these must have written instructions to cover the following:
  The objectives and extent of the work to be undertaken:

- Proposed methods to achieve the objectives.
- Identify interfaces and lines of communication between individuals and also departments.
- Specify methods of verification and checking.
- Describe methods of providing feedback.

The method by which this information is to be communicated must be made clear. Methods include:

- Handbooks
- Job specifications
- Job routine (labour returns)
- Activity checklists

Plan
Where there is to be variations to the quality system to accommodate the specific requirements of different contracts or projects, a 'Quality Plan' should specify such revisions and include the following:

- Contract Brief
- Project Phasing
- Standards and Procedures to be used (provided by a quality assessment schedule)
- Resources required for the work.

Documentation
Method of processing, storing and controlling the circulation of documentation is to be specified. Details should specify all: paperwork; computer data; register of documents; drawings, etc. There should be procedures for the review and update of documentation on a regular basis and these must be available for independent assessment.

Purchasing (including sub-letting services)
The system should specify how control of sublet work is to be conducted. If the work is substantial there must be an assessment statement as to the ability of the sub-contractor's staff and organisation. Quality Plans must be obtained from the firm and these shall form part of their contract. Their undertakings should meet the requirements of the employer's quality assurance system.

Controls
The method of monitoring, control and modification of the system should be specified. This statement should include the criteria by which the activity is measured, verified and recorded. How modification to the quality system is to be made should be specified.

Records
The method of recording information defined in the quality procedures should be specified. Records must be made systematically, indexed and continuously maintained. Records made must be available for inspection for a stipulated number of years for auditing purposes.

Review and Audit
The system should specify the method and frequency of auditing and review. Audits should cover both office and production site activities. Auditing mechanisms should accommodate both internal auditing by the organisation itself and/or its appropriate auditor, and external auditing by an independent outside organisation.

Training
There should be a statement as to the nature and method of training personnel involved in the running of the quality system. It should specify how their knowledge and skills will be updated to meet with change within the organisation. Systematic training must be adopted for all personnel. Abstracts of training philosophy and systems should be given from the company's handbook and should form a part of the quality manual.

The quality assurance system should also consider the following:

Appropriate Standards
Assessment of all quality systems is made against one standard, BS 5750, which incorporates the ISO 9000 Series of Standards. Most European countries have now adopted these standards in European Standard EN 29000. BS 5750 and ISO standards do not vary from business to business or between products or services, so considerable interpretation is needed when applying them to particular industries and situations. All certification bodies recognise these standards as does the NACCB. A clear specification of all standards should be given in the quality manual.

Development Schedule
A full set of quality procedures must be available at the time of assessment. It must be ensured that the organisation has been operating in the manner defined in their manual for a period of time long enough to demonstrate that the procedures used are effective in practice. This period must be clearly stated. The period will vary from one organisation to the next but six months is likely to be a minimum period of acceptability to the assessors. This period is required to demonstrate that the system has become 'routine' and that its effectiveness is conclusive.

'Second' Organisations
The 'hiring-in' of specialist sub-contract services is permitted providing adequate control for them exists within the quality system. The sub-letting of a major part of the organisational activity is not permitted as the quality system cannot effectively operate on a second organisation over which the main organisation may have minimal control. Any sub-contracting within the quality system must be described within the quality manual.

Use of Approval

Any organisation must not, through its quality system attempt to mislead any client, customer, or the public. Certification bodies monitor the quality system after assessment and throughout the period of registration to ensure that the correct use of documents, logo and certificate is upheld. The logo and certificate are owned by the certification body and can, therefore, be withdrawn at their discretion. Rules and regulations governing use are provided by the certification body. The intended use of logos on promotional literature and company documentation should be clearly stated in the quality manual.

The structure within which the quality system is developed and integrated can be both complex and time consuming to develop. The process involves considerable integration of management and the workforce, together with the input of consultants and the certification body. A quality system may take from a few months to several years to become fully developed and implemented to a successful outcome.

5.3 Detailing for Quality Manuals

This section outlines a fictitious but typical structure, content and level of detail required for information being presented in the Quality Manual. It gives details of content in sections for: Policy; Organisation; and Assessment and Review, and also the typical presentation of diagrammatic information, where this is included to supplement the text.

The content of the quality manual should be concerned specifically with the company's policy, structure, organisation and procedures relating to quality and its assurance. More general detail of the company's organisation and activity can be provided in the form of a 'Company Handbook' of standard procedures which can accompany the quality manual when applying for the assessment and certification of the quality system.

Pro-Forma Notes for Quality Manuals

Statement by the Company Chairman

This statement should be accompanied by company letterhead and logo and be official in its presentation and take a form similar to that which follows:

> The corporate philosophy, aim and purpose of our Company is to provide products, goods and services of outstanding quality and with first class efficiency. Quality, performance, reliability and safety are all

essential to the continued success and prosperity of our company and the well being of the industry which we serve. Responsibility for quality achievement falls upon every employee within our company. Each employee is responsible to their direct superior who is in turn accountable to the Company's Quality Assurance Manager and Quality Assurance Management Group. This Group will co-ordinate and control all aspects of quality assurance within the organisation to ensure our clients and customers of our commitment to the pursuit of quality.

Signature of the Chairman

Quality Standards

Specified Standards – The Company's quality system should comply with the appropriate part of BS 5750: Quality System and, where appropriate, International Standards such as ISO 9000 Series. Quality systems developed for divisional groups within the control of main organisation should also comply with BS 5750. Sub-contractors must also operate a recognised quality system to BS 5750 and all suppliers should be procured from a BSI approved list of manufacturers and suppliers.

Policy

(i) *Scope*

This section (Policy) describes the Company's organisational policy in respect of quality achievement within the aims and objectives outlined in the Chairman's Statement.

(ii) *Introduction*

In the context of the Quality Manual, Quality Assurance describes all organisational activities and procedures for ensuring and demonstrating that the required standards of quality are achieved in all organisational activities.

(iii) *Corporate Responsibility*

The Company's Quality Assurance Management Group will assume responsibility for the development and continued implementation of the Company's quality assurance system. This Group will be accountable to the Company's Chief Quality Assurance Director who is in turn accountable to the Board of Director's Chief Executive or Company Secretary.

(iv) *Range of Responsibilities*

Each Division or sub-section of the Company is responsible for assuring the quality of its services and products. As the company has a wide range of activities in the areas of supplying products, materials and services to industry, it is not sufficient to maintain a single quality system which is common to all sub-groups. For this reason, each division, within the framework of the Company is required to develop a quality assurance system to meet the needs of its own activities. This system should include policy, (in the context of the Company's philosophy or objectives), organisation, delegation of authority and procedures for monitoring, control, audit and review. These procedures should include the Company's own activity and those of its sub-contractors and suppliers.

(v) *Management Responsibility*
All employees with managerial responsibilities are to be responsible for the achievement of quality. Management is required to ensure that: all work carried out under their control complies with the required specification including the work undertaken by sub-contractors and items ·procured from suppliers; that staff directly under their control are familiar with the Company's organisational policy in quality assurance and also with any particular requirements of the quality system relevant to their work; and that all staff assigned to projects are suitably qualified, capable and experienced in their roles and duties to perform their function to the required standard of quality or workmanship.

(vi) *Responsibilities of Individuals*   Every employee with the Company has a duty of care and responsibility to the organisation to ensure that their work complies with the required standard of performance.

(vii)*Implementation of Policy*
The Company's Quality Assurance Policy is implemented within the remit of the Quality Assurance Management Group through the organisation and procedures which are described in the Quality Manual.

Organisation

(i) *The Company*
The principal activities within the Company and its associated sub-divisions are:
  • Building and Construction
  • Civil Engineering Works
  • Housebuilding
  • Building Maintenance and Refurbishment

(ii) *The Company Quality Assurance Management Group*
The Company's Quality Assurance Management Group provides the organisation with a corporate approach to quality assurance as described. Responsibility for this Group is assumed by the Quality Assurance Manager (named personally).
   The function of the Quality Assurance Management Group includes:
  • Preparation of quality assurance documentation
  • Initiating and developing the quality system

- Providing advice on quality systems to the Company and its sub-divisions
- Audit analysis and review of the Company's activities
- Independent assessment of quality assurance organisation from the Company's own organisational structure
- Representation to clients
- Liaison with the public
- Co-ordination of quality assurance training and staff development relevant to the operation of the quality system

(iii) *Structure of the Quality Assurance Management Group*

The structure and organisation of the Company's Quality Assurance Management Group and its relationship to the Company's framework is illustrated in Figure 5.2.

(iv) *Divisional Responsibilities*

All contracts require Quality Manuals and Plans describing the systems needed to ensure compliance with the requirements and specifications for the project. These manuals and plans should follow the format of the Company Manual. Small projects do not require a quality manual but should be described by a Quality Plan. Contract or Project Managers are responsible for the development of such plans and to ensure that their content is implemented. The requirements for such a divisional structure is shown in Figure 5.3.

Review and Assessment

(i) *Definition of Activities*

Account of the Company's quality assurance activities will be undertaken through formalised and documented audit and review.

The definition of activities are as follows:

'Audit': An independent examination of the quality system to provide assurance to senior management that the quality assurance system is effective in implementation and in being fully complied with.

'Review': An examination by the Company's Quality Assurance Management Group to assess the effectiveness of the Company's quality system and implement change if it is not to the required standards or operating procedures.

(ii) *Audit*

Audits carried out within the Organisation include:

- Audit of quality system by Company Quality Assurance Management Group.
- Audit of Divisions by the Quality Assurance Management Group.
- Audit of Departments by Divisional Quality Assurance Manager.
- Audit of Projects by Divisional Quality Assurance Manager.
- Audit of smaller contracts by Contract/ Project Manager.
- Audit of suppliers by Purchasing Department.

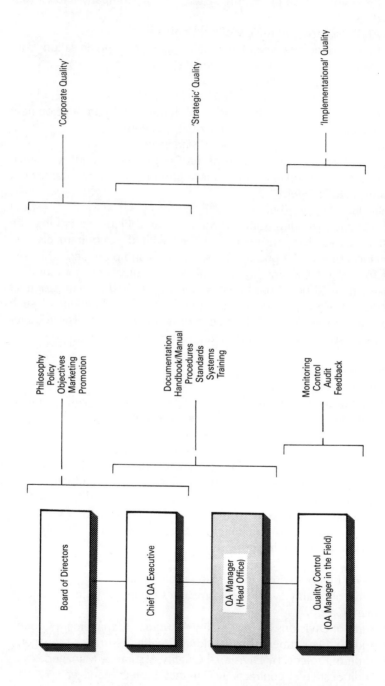

Figure 5.1   The key role of the Quality Assurance Manager within the structure of the Quality System

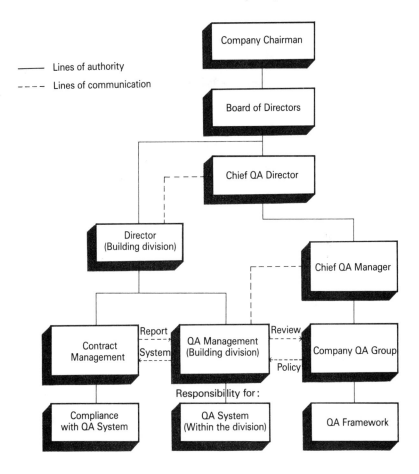

Figure 5.2 Structure and organisation of a company's assurance management group activities

(iii) *Audit Objectives*

The purpose of audits are to determine that the policy, aims and objectives are being met and being effectively implemented by all divisions and projects within the framework of the Company.

(iv) *Audit Procedure*

The Company Quality Assurance Management Group's Chief Quality Assurance Manager will arrange all audits. Arrangement will be made verbally and confirmed to all parties in writing. He should prepare: a check list of principal aspects of the organisation to be audited; an agenda of key events; and gather all documentation necessary to undertake the audit.

The auditee will make arrangements for appropriate personnel and documentation to be available to respond to the requirements of the

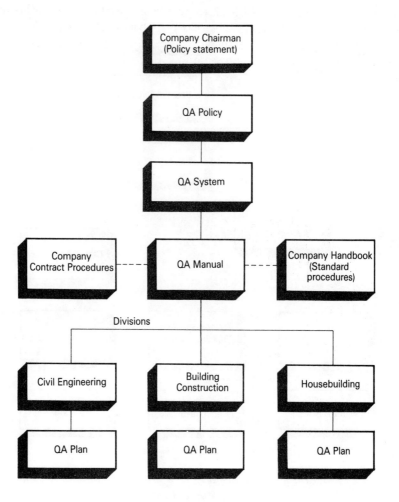

Figure 5.3 Quality assurance requirement for divisional structure within an organisation

check list and in accordance with the auditing programme.

Auditors will examine the documentation and inform the auditees of any queries arising from their examination before the formal audit.

Audits will commence with a meeting, then proceed by examination of the documentation, system, review of the system in progress and conclude with a verbal report.

A written report will follow, within a specified time, and list the findings of the audit and specify any non-conformances.

Corrective action should be taken within a specified time, following receipt of the audit report. Copies of the report will be available to the Company's Chief Quality Assurance Executive, the Chief Quality Assurance Manager and Divisional Quality Assurance Manager.

6 Quality Assurance: its Application in Construction

6.1 Awareness within the Construction Industry

Until recent years, the formal application of quality assurance within the UK construction industry had been limited to a small number of complex, high-value and high-risk projects, primarily in the nuclear and power industries and in off-shore and petrochemical engineering where reliability and safety are paramount. Quality Assurance in other sectors of construction industry has generally been confined to the manufacture of materials and component products where quality assurance has been implemented for many years, although Government-led pressure is now demanding the adoption of quality assurance in many other sectors of construction.

The factor that sets construction apart from manufacturing is that, in most applications of quality assurance, there is no clearly identifiable end product. It is difficult, if not impossible to quantify the quality of the design process or the quality of site management and it is this vagueness that presents the major barrier to the widespread implementation of quality assurance at this time.

Many major building contractors see quality assurance as a logical and progressive step beyond merely good management practices. If they undertake regular public sector work then there are obvious benefits in aiding the client's rigorous pre-tender assessment processes. Few contractors however, operate officially registered third-party quality assurance systems and are tending to develop in-house procedures to regulate their own activities.

The main thrust of sector based professional interest in quality assurance has come from the building services engineering sector of the construction industry which has strong links with general contracting organisations. A number of services engineering companies have developed quality assurance schemes to meet their own particular requirements. This has given rise to the Building Engineering Services Certification Authority (BESCA), the services' own approval agency, which is accredited by the NACCB.

In other aspects of engineering such as, the structural field for example, there has been minimal pressure to pursue quality assurance certification. Nonetheless, many practices are viewing the increasing client/customer expectations of quality as an impetus to appraise and improve their systems and services in consultation with the British Standards Institution and other certification bodies.

In the design sector it is more difficult to assure a client of the level of expertise available other than through the recognition of past performance. There are few formalised quality assurance systems in architectural practices and awareness and interest has generally been slow to emerge. The architectural profession have declared an interest in quality assurance but hold the general view that quality is an inherent aspect of their professional qualification and standing that no specific quality assurance measures or systems are, as yet, required. Quality assurance differs from professional qualification in that QA is a mark of the organisation's ability to provide a good quality service, and therefore should not be confused with the standing of the individual. This may change as more clients become interested in quality assurance and may procure design consultants along with other professional services on the basis of them having full certification.

In the fields of quantity surveying, building surveying and general practice surveying, the awareness, recognition and acceptance of quality assurance has also been slow to develop. Like the architectural profession, the surveying profession is overseen by its own professional body, the Royal Institution of Chartered Surveyors (RICS), who have more recently perceived the need to issue guidance documents for quality assurance throughout their profession. Some quantity surveying practices have however, taken the initiative. One such practice, believes that third-party certification will, in the future, become a prerequisite and mandatory aspect of contractual arrangements with the client. As such, they have taken the step of achieving registration of their company wide system with the BSI/QAS scheme.

As quality assurance becomes better understood within the construction industry, the requirement for independent recognition and certification of quality assurance systems will increase. Many of the larger UK public and private sector clients are encouraging the use of third-party quality assurance services. The Property Services Agency (PSA) has, for some years, invoked formal quality assurance through its Method of Building Scheme which controls carefully the selection of material and component suppliers and use of products on its Government contracts. This philosophy is likely to continue as the PSA procures its architectural, consultancy and contracting services only from BS 5750 registered sources, a pattern likely to continue well into the 1990s. Similarly other larger bodies including the Department of Transport, British Rail and British Telecom are all poised to implement policies requiring the use of independently assessed services on their contracts. Such endeavours will certainly place a new emphasis upon quality assurance, particularly if quality assurance becomes an integral part of contract documentation and forms of contract.

6.2 The Effect of Procurement System

Quality systems, for which BS 5750 is the basic standard, provides a general management framework for defining and meeting the requirements of products or services. Where the design and production processes come under a single point of control and responsibility, as in the manufacturing industry, the application of a quality assurance system is relatively straight-forward. The standards and guidelines of BS 5750 lend themselves to the close control of quality in such an application. Within the building and construction industry, one of the most effective applications of quality assurance systems is to be seen in the production of materials and components which in many ways follows very closely the same pattern as manufacturing.

There should, in principle, be little difficulty in transferring the basic concepts of manufacturing industry to the building process where, for example, the design-build method of building procurement is used, as the design and production aspects are integrated and come within the remit of a single source of responsibility. The 'traditional' form of building procure-ment however, where each building professional acts in an independent role, presents distinct problems for the implementation of quality assurance. The problems arise through the multiplicity of design and production responsibilities involved in the contractual form and moreover, from the lack of integration which occurs through the traditional separation of the various parties. The individual parts of BS 5750 are not applicable to construction 'across-the-board' because the construction industry and its professional parties tend to utilise a conglomeration of 'performance', 'product' and 'method' specification in the procurement process, and, of course, a building is not a product or service in the conventional sense.

Where there is a 'project' emphasis to the structure of the contract however, BS 5750 becomes much more meaningful and directly applicable since 'quality management' becomes an inherent and important aspect in the procurement structure and organisation. In this approach, the Project Manager can assume total responsibility and ensures that each party is upholding its own system and commitment to quality. Effectively this does mean however, that 'total quality assurance' in construction can only be achieved successfully if a particular approach to procurement, such as project management or design-build, is used. Certification bodies recognise the importance of such non-traditional forms of building procurement and have developed specific quality assurance systems to meet these require-ments. Such applications follow in this chapter.

Although non-traditional procurement techniques are increasing in popu-larity and use, a greater proportion of construction work continues to be commissioned under the traditional procurement approach. Given this limitation, a framework across the construction industry can only come from the integration of the individual quality systems implemented by the various professional parties within their own sectors of the construction industry. This requires that each individual party, whether it be designer, contractor, supplier, or consultant, considers its own philosophy towards quality. Each organisation is required to develop its own quality system, obtain certification and maintain and uphold its own beliefs and commitment to the pursuit of quality.

6.3 Quality Assurance in 'Traditional' Building Procurement

The overall aim of any construction project is:

the design and construction of a building or structure to meet the specific requirements of the client.

Quality in building construction can be achieved only if the many and diverse factors contributing to efficient and effective building makes it possible and moreover, that the various professionals involved throughout the total building process perceive the need for quality with integrity and commitment. Systematic methods of quality assurance and management creates a discipline on the outline procurement and production process. Quality achievement requires the effective functioning and integration of, informed clients, empathic designers, knowledgeable suppliers of contract documentation and responsive contractors. Together, these professionals must address a number of fundamental issues:

- The client clearly defining the requirements for the building.
- The designer and other consultants accurately interpreting the client's needs and form of building and specifying materials and components to meet these requirements.
- The building contractor translating the design concept and physically creating the building to meet the client's needs, as perceived by the designer.

The modern building process is, in many ways, detailed and complex, with a wide range of criteria to be satisfied, and the many quite separate professional parties needing to be brought together. In any construction project, there is some risk and some chance element, however small, that difficulties will arise and that something will go wrong. Whilst problems can never be eliminated, difficulty can be reduced through the careful attention to the roles and responsibilities that the client and each professional party assumes during the project.

Problems of Implementation

Problems experienced in the course of implementing quality assurance within traditionally procured construction manifest primarily at two points in the total building process: at the interface between the client and designer and the interface between the designer and the contractor.

The principal reasons for the project failing to meet with the client's requirements include the following:

- Client and designer may misunderstand or not agree upon the details of the brief with regard to the building's purpose, performance or appearance. The brief must be precise and convey the client's genuine needs. Any discrepancies will adversely affect communication, co-ordination and ultimately the level of quality achievable.
- The designer may perceive quality in a different way to the client. Quality levels must be defined clearly by the client and the designer's brief be tightly controlled so that the designer does not consciously resist or inadvertently ignore the wishes of the client.
- The designer may, through misunderstanding or ignorance, misinterpret the objectives regarding quality specified by the client. The designer can 'design-in' his own interpretation of quality where a brief is not specific enough.
- The designer may design the building that cannot be built to the required levels of quality through incompatibility between design and the construction's buildability. Designers may develop concepts that simply cannot be built without cost and time consuming modifications and re-drafts.
- The construction may be insufficiently financed to build to the levels of quality desired by the client. Failure to balance cost with quality at the brief and design stages can lead to a reduction in the level of quality achieved during construction through re-specification of some items to an inferior standard.
- There may be an insufficient time schedule to build to the desired quality level. Time constraints must be carefully balanced with cost and quality at the design stage to avoid overrun on works or risking the project and invoking poor construction quality.
- The contractor may fail to understand what the desired standards of quality are. If the design stage is completed before appointing the contractor, as in the case of 'traditional' procurement, there is no opportunity for the designer to effectively communicate desired quality levels to the contractor first hand. The designer has to rely on the clarity of drawings and specifications and upon the contractor's correct

interpretation of these when construction proceeds. The builder does not always know the 'real' standards before commencing work on site.

- The contractor may not be able to build to the desired quality standards due to a mis-match between the design and construction systems. The contractor may not be sufficiently competent to undertake the task or may misunderstand the requirements through inadequate communication of the design concept.
- Design requirements may not be communicated effectively to the operative at the workplace. The designer must convey the design intention simply and effectively to the contractor, whose management and site supervisory staff must convey equally as accurately to the operative at the workplace.
- Discrepancies occur between the anticipated and actual life cycle cost of the project. This can result from an inadequate brief making the building too expensive to up-keep and maintain.

Quality Assurance Application

Quality assurance must be actively employed throughout the total building process from initial briefing and conceptual design, through the assembly process to completion of construction. Throughout the process, it is essential that clear communication is encouraged, in particular at the interfaces of project responsibility and control.

Quality assurance can be applied at the three main stages of the building process:

- Briefing, design and planning
- Manufacture and supply of materials, components and products
- Construction

A fourth stage, that of commissioning, handover and the use of the building is equally as important as the preceding aspects of construction and therefore it should not be ignored, nor under-estimated. As a total concept in construction, quality assurance certainly has an important part to play in this final stage through inspection and testing, instructing the occupier on the use and running of the building and in assessing maintenance and upkeep requirements.

The standards of performance and quality desired should be determined between the client and the designer and, where possible, the contractor at the feasibility design briefing and project appraisal planning stage. A Quality Plan specific to the requirements of the project can be developed. This plan is the document that sets out the quality assurance system arrangements and procedures that will be used to meet the requirements of the contract and specification set by the client. The designer will produce the quality plan, in which the quality standards will be specified. The contractor will liaise with the designer in working to these standards.

Similarly, the designer will issue a quality plan to any supplier he employs directly. If the supplier, or sub-contractor, is to be employed by the contractor, the designer will issue a quality plan through the contractor or supervise the issuing of a quality plan from the contractor directly. In either event, the quality plan should ensure that the client's requirements are clearly specified.

The supplier of goods or services should give evidence of its capability to ensure quality in its products, determine that they are fit for the intended purpose and should agree to contract time scales within which their goods or services will be provided. The acceptance of the quality plan issued to them should be a condition on which the order is placed. If the work that the supplier will undertake is considerable, they should demonstrate the ability to produce their own quality plans in respect of their commitment and see this conforms to the employer's overall quality plan. The capability of a supplier can be assessed by the designer or contractor in several ways: by investigation and evaluation of the supplier's current and recent activities; through experience and past involvement; and the supplier's involvement with a recognised third-party certification scheme.

The progress of orders placed for the procurement of goods and materials should be followed by the purchaser through all stages of manufacture and delivery, and in the case of services, close liaison maintained before their arrival on site. Close control in this way allows modifications to be made to the overall quality plan to ensure the continuity of quality and performance. The building contractor should be fully aware of the current status of all materials, goods and services procured both by himself and the designer. Quality products or services serve little use if they fail to arrive on site to the agreed schedule or to the appointed place. For reasons such as these, the planning of products and services has become an important aspect of quality assurance systems and is reflected extensively in BS 5750: Quality Systems.

The quality assurance system should arrange that adequate precautions are taken to ensure that materials and components delivered to site are adequately protected against weathering and damage, are handled carefully during delivery and stored on site safely until used in the works. Each delivery should be identifiable and easily traced back to the supplier in the event of defect. The contractor should be aware of the material's or product's handling, storage and assembly requirements and also be fully informed of any limitations in its use.

On-site quality assurance management, should ideally, be independent of the influences of the building programme and have the necessary authority to ensure that the necessary actions are taken as and when required to maintain quality. In practice, the on-site responsibility for quality assurance often falls upon the architect's site representative, the clerk of works, but, despite undoubted experience the clerk of works does not always have the required authority nor time to fully meet the task. It is, therefore, advisable for the client to appoint an independent person to safeguard his interests for performance and quality, a quality assurance consultant. Likewise, the building contractor should not rely upon the site agent or foreman to assume the onerous responsibility for quality assurance on site and, therefore, appoint a quality assurance manager to assume full responsibility for its achievement.

The quality assurance manager, for the designer and contractor respectively, should monitor and inspect the conformity of all services to the specification. They should also ensure that materials and products also meet the specification and that the storage and installation is to the client's requirement or manufacturer's recommendation and are fixed in accordance with building regulations and to sound construction practice. Such procedures should form the major part of any Quality System to ensure that quality assurance of products or services becomes a commitment throughout the total building process.

Client Involvement

The client can directly promote quality assurance in a project by:

(i) Procuring design from a reputable architectural practice, ideally one that operates a quality assurance system within its organisation. (As no design practice has at the time of writing this book, obtained third-party certification, quality assurance systems are likely to be only first or second party in type and scope.)

(ii) Procuring construction work from a reputable building contracting company, whose organisation operates a registered quality assurance system.

(iii) Appointing consultants who procure materials, components and products from the British Standards Institution – BSI List of Certified Suppliers and Manufacturers.

Whether goods or professional services are procured from certificated or non-registered organisations, the client should express commitment for quality and set quality standards within the project structure, organisation and documentation, and maintain this commitment throughout the overall management of the project. The appointment of registered construction services and the use of certified products will not guarantee project success but will go some way to assuring the client of a high quality input to the project which should certainly contribute to greater potential for project success. The client must incorporate quality assurance throughout his early involvement in the project, during the initial formulation of the brief, in the procurement of consultant services and advisors, for the project appraisal and outline feasibility stage.

Clients giving due consideration to quality should address the following aspects of BS 5750: Quality Systems in the procurement of their projects:

- *Responsibility for quality*   the client should appoint one person to represent his interest. (This person may be the architect or an independent 'Quality Assurance Management Consultant'), for example an in-house quality assurance manager.
- *Quality System*   the client should develop a 'project' quality assurance system to a formalised structure and organisation, within the remit of the client's quality representative. This may involve a 'quality assurance team' under the direction of the QA Manager.
- *Planning*   the project quality assurance system should appreciate the role of design, manufacture, construction, and other construction services and incorporate their operations within the project quality system to maintain consistency in quality, particularly to the project in hand.
- *Documentation*   the system must formalise the client's requirements for documentation, instructions and records from all contractual parties. Documentation will define the criteria for the project and set performance and quality levels relative to other key project parameters such as time and cost.
- *Control*   the client should consider how the system will implement inspection, measurement, testing and control in association with other contractual parties.
- *Sub-contractors and Suppliers*   the client should formalise his dealings with all parties supplying goods or services, when procured directly by the client.
- *Protection of Quality*   the client should consider how quality will be protected and preserved within completed works whilst construction proceeds, and during other aspects of the works such as material storage on site. Consideration must also be given to the security of both the works in progress and to finished works.

The levels of quality desired by the client are usually associated with those expressed in the client's brief. The client does however, have a duty of care and responsibility for quality assurance throughout the project in the role of employer to the designer and in dealing with the main contractor, sub-contractors and suppliers. The client can have a major effect upon quality in the following aspects of his work:

- In the Project Definition: development of the brief, consideration of economics, description of the physical requirements;
- In Setting Constraints: performance, time, cost

and in consultation with the designer, the following aspects:

- In Communication and Documentation: specification, schedules, drawings, meetings, records
- In Controlling: inspection, testing, correction, improving, auditing, review and assessment

The client plays his most important part in the achievement of quality before the project even reaches the drawing board. The time and cost constraints specified by the client must be realistic and sufficient to give the client the standards of quality expected. During formulation of the brief, the design, and the construction process, the client must be aware of the performance, cost and time consequences upon the decisions he or his professional advisor takes. In appreciation of this, the client must be well guided by the professionals who represent his interests throughout the total building process.

The client must be able to make decisions from the outset. Not all clients have sufficient knowledge, skills or inclination to maintain very close interest in their projects, but quality is likely to be better where the client has the ability to make quick and positive strategic decisions during the design and construction stages and, moreover, if he becomes actively involved in the processes. Difficulties can arise at the initial briefing stage if quality standards have not been fully thought through. The client therefore, must give adequate consideration to, and unequivocally specify, the building's purpose, expected performance (both functional and technical); its desired appearance, its overall cost and cash flow constraints, and time schedules needed for completion. It is essential that the client defines the standard of quality required when the brief is prepared and that the designer explains the range of standards that are achievable given the practical constraints of time and cost.

It should be noted that the architect is not always the best person to represent the client concerning aspects of quality as he is directly responsible for the design and may see the quality of the building in a limited, even biased way. For this reason the client may decide to appoint an independent advisor, a quality assurance management consultant, to safeguard his requirements where quality is concerned. This may also serve to reduce potential conflict for the architect.

For effective quality assurance from the client's viewpoint, the client should undertake the following practical steps:

- Be involved from the outset of the project.
- Take an interest throughout the total building process.

- Take a prominent role in briefing, design, production of documentation and project control, where appropriate.
- Clearly specify project requirements for quality and quality assurance to the designer.
- Appoint a Quality Assurance Manager, (outside the role of the architect) for independent and objective guidance in the aspect of project quality.
- Develop and implement a formal quality assurance system for the project in which all parties must contribute as a requirement of their contract.
- Show commitment to the achievement of quality in all aspects of the project.

The Role of Design

Design consultants can influence quality assurance at a number of key areas within their remit. During the 'pre-construction' stage these include the following:

- Outline or conceptual design
- Cost planning
- Final design
- Appointment of main contractor
- Procurement of suppliers and sub-contractors

Outline Design
A successful outline design relies heavily upon close integration and open communication between the client, the design team and the various project consultants. The main aim at this stage is to introduce desired aspects of quality and performance for assessment within the context of external variables, such as building regulation or planning restrictions, and other project variables such as time, cost and available resources. The realistic balance between quality and constraints should be determined at this stage such that an outline quality plan for the project can be suggested.

Cost Planning
During formulation of the cost plan, the cost implications of quality assurance must be considered. As in the outline design stage, a realistic balance must be maintained between the level of quality desired and the level achievable given the influence of other project variables. The client must, therefore, be made aware of expected quality levels arising from the suggested design and specification so that the client's understanding of value for money is realistically determined.

Final Design
The final design stage integrates and consolidates the outline design and financial plan through the production of the detailed design. During the final design stage the philosophy towards quality can be clearly established through developing the outline quality plan and consideration given as to how quality assurance will become an integral and functional part of the construction phase of the total building process.

Appointment of Main Contractor
Quality assurance is enhanced significantly where a contractor takes an active part in the design process. Within the 'traditional' approach the separation of design team from construction team invariably means that the contractor's potential contribution at the pre-construction stage is never realised. In simple terms, the earlier the main contractor is appointed and incorporated into the 'building team' the greater the likelihood of project success. Quality assurance is an important consideration for the design team, with reference to and guidance of the client in the appointment of the main contractor. The following should be considered:

- Contractors considered for the project should have recognised experience in using quality systems.
- The contractor should be sufficiently capable of modifying his system to the requirements of the client's and design team's project quality philosophy.
- The requirements for a modification to the quality system should be clearly stated in the tender documents.
- Specific requirements for suppliers and sub-contractors should also be stated in the project documentation.
- The level of quality system required should be clearly specified, for example first party, second-party or full third-party certification to BS 5750, together with any quality schedules pertinent to specific activity or specialisms required.

Procurement of Suppliers and Sub-Contractors
The same criteria that apply to the selection and appointment of the main contractor should be extended to the acquisition of materials and components and other construction services. As many products, materials and components suppliers are already subjected to third-party quality assurance schemes under BS 5750 the design consultants should make it a rule only to procure form such recognised sources. In the same way the appointment of sub-contractors should follow only when it has been seen that they can meet the requirements for quality specified in BS 5750.

The Requirement of the Design Consultant

BS 5750: Quality Systems provides a systematic framework for the organisation of consultant design and its management. Again, manufacturing
forms the basis for its description but key aspects can be related to design
activities of the project situation within construction.

The outline design guidance of BS 5750 requires that any design
consultant undertakes to provide to the client the following:

- Provide a design and development programme for each project.
- Provide a code of its own design practices and procedures.
- Investigate new technologies as an integral aspect of design tasks.
- Identify and control design interfaces between contractual parties.
- Maintain project documentation (including drawings, specifications,
  procedures and instructions).
- Control tolerances (and appreciate the reality of site tolerances during
  the formulation of the design concept).
- Consider all statutory requirements (including health and safety).
- Evaluate new materials and components for potential use.
- Control reliability and appreciate value for money considerations.
- Establish design review procedures to assess progress and consider
  problems.
- Provide feedback from previous design and construction experiences
  for assessment and learning.

The client should, where possible, procure design services from architectural practices who can clearly demonstrate their commitment to quality
assurance. Although design practices have yet to become associated with
third-party certification schemes to BS 5750, many adopt self-regulating
quality systems that can meet the requirements of more discerning clients.

Certification bodies such as the British Standards Institution, Yarsley and
Lloyds, have guidelines for design practices to the standards of BS 5750:
Quality Systems or have documentation that can be adapted to such
design orientations. Such guidelines (Quality Assurance Supplements –
QAS) require the design practice to meet the same basic organisational
requirements for quality assurance that any other organisation seeking
registration of their quality system must meet, but in addition there are
specific requirements for design services.

For example, an architectural practice will have to, meet the requirements
of, provide documentation for, and implement the following:

- Quality Policy
- Organisation
- Quality System
- Planning
- Auditing
- Recording

- Control
- Review
- Training

These represent the basic requirements of any registered quality assurance system. In addition to these requirements, specific requirements must be addressed. These include the following:

Contract Plans

Plans of work for each project should include all documents of the work to be undertaken. This should cover the total building process from client's brief to completion and must include as a minimum requirement the following:

(i) The method of ensuring adequate planning of all aspects of the work.
(ii) The project structure indicating functions to be carried out as specified by the client. The plans should identify all personnel involved and resources needed to be brought in for the particular project. Responsibilities and authority of staff should be clearly defined.
(iii) A project programme showing phases of the project including responsibilities during tender and also site activity should be summarised and described.
(iv) Project procedures which are additional to corporate operational procedures should be specified and described.
(v) Plans should identify all design reviews including the method of obtaining the client's and statutory approvals.

Contract Review

Subsequent to the award of a contract, the design practice should have a system for reviewing the client's requirements and should consider the following:

(i) Specific contract requirements including all regulations and statutory requirements.
(ii) Aspects of technological, financial and quality assurance documentation.

Design Control

The design should reference the standards against which it has been prepared, that is British Standards: codes of practice, Building Regulations, statutory requirements and client's specification. The design practice should prepare and describe detailed procedures for control, including:

(i) Method for preparing and checking drawings, schedules and bills of quantities.
(ii) Method for controlling the work of all other parties.
(iii) Method for issuing and controlling project documentation and information.

Sub-contract Works

Where a design practice procures sub-contract services directly for design contribution or selects the contractor for the client, then an objective basis for selection must be adopted and any work procured in this way must satisfy the requirements of BS 5750, to ensure the continuity of quality throughout the design process.

The overall responsibility for ensuring that the appropriate quality is achieved is assumed, under the traditional form of building procurement, by the architect, unless otherwise arranged. Where formalised quality assurance to the requirements of BS 5750 is adopted the achievement of quality should not be a problem. Until quality assurance becomes more widespread in application however, the construction industry will have to rely upon less formal arrangements.

The project architect plays a key role in the achievement of quality assurance. The design phase must take account of the process by which a building is constructed in addition to what the building will look like when completed. The designer must appreciate the complexities of the building process and materials technology used in the production process if good quality is to be achieved. The designer has a duty, not only to interpret and present the client's brief, specification and documentation, but be committed to maintaining and controlling the design during its execution on site. The architect essentially assumes the role of: designer; specifier; and production supervisor.

For effective quality assurance the design team's project architect should undertake the following aspects all of which really represent little more than simply good design management:

- The careful preparation of design details, specification, drawings and documentation.
- To clearly specify the desired standards of quality and performance required by the client in all contract documentation.
- Specify materials and components that are BSI approved.
- Design the building with empathy for the construction stage to assure buildability.
- Provide project information (drawings, schedules, specifications, and so on) concerning quality speedily and continually throughout the project.
- Inspect all materials and components delivered to site and used in the works.
- Testing of materials for quality levels when required.
- Informing the contractor of work not meeting the contract requirements without delay.
- Take action to rectify discrepancies in drawings, documents and instructions promptly.

- Provide quick response to requests for information to describe and detail design changes.
- Provide feedback to both the client and the contractor on progress and performance.
- Compile records for analysis and assessment by all parties for use in procuring future contracts.

Specification and Procurement of Materials and Components

The architect is responsible for and should ensure that any material, product or component used in the work is of a suitable type and quality for the intended purpose and conditions, adequately prepared or mixed, and applied so as to perform its intended function. Where it is possible, the architect should ensure that the materials, goods or components are specified and procured from a reputable source that operates its own quality assurance system, that complies with BS 5750 and is listed on the BSI Register of Firms of Assessed Capability. Where this is not possible, the architect should ensure that the suppliers meet with the advice of the Department of the Environment (DOE): Approved Document No 7 and also that one or more of the following points be considered as appropriate:

British Standards
The product should conform to the standard which is appropriate to the purpose and to the conditions under which the product shall be used.

Agrement Certificate
The material should conform to assessment by the British Board of Agrement (BBA); that a certificate has been issued in this respect; and that the product or material is used in accordance with the terms specified by the Agrement Certificate.

Independent Certification
The material or product meets with the requirements of an independent certification body such as the 'Kitemark Scheme' of the BSI or the 'Testguard' product conformity certificate issued by Yarsley, or that the item has been tested by a recognised testing body under NATLAS.

Building Regulations
The material or product meets with the relevant requirements of the Building Regulations.
   Where the architect is responsible directly or supervises the main contractor in procuring supplies, a sound approach would take in the following:

- To provide a clear description of the material or component required in the project documentation and specification.

- To ensure that all requirements are specified in enquiries and orders placed with suppliers.
- To provide a method of checking materials and components delivered to site.
- Where materials and components require inspection offsite, during say prefabrication, to make provision for this.
- That suppliers are given sufficient information about the project, such as time schedules, to ensure they respond to the constraints of the project.

Procurement of Sub-contract Work

The architect assumes responsibility for any specialist works undertaken for the project that are procured by him directly. As in the case of ensuring quality of materials and components, so the architect must ensure that the desired levels of quality in workmanship have been met. The architect should therefore only procure services from a sub-contractor that meets one or more of the following conditions:

- Operates a quality assurance scheme to BS 5750, has certification from the British Standards Institution, and appears on a recognised list of assessed services issued by the BSI.
- Is registered with an independent quality assurance scheme such as Yarsley or Lloyds, has full certification, and appears on a recognised list of assessed services issued by these bodies.
- Offers a first- or second-party quality assurance system which can be assessed by the architect and has the ability to meet with all the requirements as specified by the client or architect.

The architect should for sub-contractors, as for suppliers, ensure that a high degree of control is maintained over:

- The selection of sub-contractor
- The issue of orders
- The monitoring of performance and quality achieved
- That records are kept for future projects
- That complete information is provided to sub-contractors, that inspections are made and that project influences are fully considered.

The Role of the Contractor

In recent years, the BSI Quality Assurance Service: Building Construction Group, along with independent certification bodies have been developing schemes to cover most areas of the production aspects of the construction process. The BSI scheme for example, is based upon the system for Registration of Firms of Assessed Capability. Schemes are designed on a modular yet integrated basis to accommodate the various services that may be offered by an organisation and to ensure that each scheme is complementary. Within construction contracting there are specifically defined categories of organisation, building construction, civil engineering contracting, building services, and so on. To ensure consistency in procedures across the various quality assurance systems, 'Quality Assurance Schedules' (QAS) are produced with different schedules to meet the requirements of each scheme.

These schedules serve first, as a guide to potential specifiers (architects) or employers (clients) of contracting companies within the scheme and give details of the range and scope of procedures within the quality assurance system. Secondly, they set desired quality system practices to which each company must adhere under the BSI System of Registration for Firms of Assessed Capability. Thirdly, the schedules complement the requirements of the BS 5750: Quality Systems and instructions of the various third-party certification schemes.

Like all other aspects of BS 5750, some interpretation is required to translate the manufacturing orientated concepts to application within the construction industry, although this is not an insurmountable problem. Essentially a contractor must demonstrate to the client that the product or service he is providing for the contract meets with the specified requirements. This can be achieved through a fully registered third party quality assurance system or a self-regulating in-house scheme that meets with the requirements specified by the client and architect.

Where a full third-party certificated quality assurance system is required, the building contractor must show that he has a fully operational quality system to the requirements of BS 5750: Quality Systems, and that this has been successfully implemented and certification obtaining in the usual way. The requirements for such a scheme are described in Chapter 5.

The contractor can play a significant role in quality achievement at two main stages in the construction process:

1 At the 'Pre-construction' stage
2 At the 'Construction' stage

Once the contract has been let and there is a commitment to proceed with the construction phase, the contractor must organise the production process. As an integral aspect of this task is the structure and organisation of quality assurance management. The immediate problem that the contractor faces is the limitation of time. There must be rapid organisation of the head office and project site teams and consideration for quality achievement must form part of the strategies adopted within the following aspects of the project:

- Selection of management team for the project.
- Meeting to discuss the contract with the client and design team.

- Assessment of the true implications of the contract, specification and drawings.
- Selection of sub-contractors and liaison with sub-contractors and suppliers procured directly by the client.
- Procurement of project site supervisors and foremen.
- Arrangement of financial guarantees and insurances.
- Systems for payment of salaries and production bonuses.
- Selection of labour, plant and materials.
- Arranging site structure and organisation.

The management arrangement for undertaking these tasks is complex and far from easy to accomplish. The achievement of quality assurance must be considered at each stage of the construction process and at all management and operational levels. Quality on site or workmanship ultimately depends upon the quality of the operative carrying out the work, but also depends considerably upon many and varied factors such as communication of the design, and the quality of supervision, leadership and motivation on site. Quality for the contractor on-site is crucially concerned with three variables:

1 Performance in relation to
2 Time
3 Cost

Under the traditional form of building contract, as described in the JCT Standard Form, the building contractor is responsible for the execution of good quality work in accordance with the specification. This responsibility falls upon the site manager, agent or foreman, supervised by the project architect or clerk of works. Under BS 5750, the contractor is responsible for appointing a Quality Assurance Manager to assume the role and responsibility for overseeing all aspects of quality assurance. The structure and organisation of the contractor's quality system should present the formal procedures under which the desired levels of quality specified in the contract can be achieved.

The main reasons for failure to produce the standard of quality required onsite include inadequacies in:

- the understanding of quality levels required
- the definition of site duties to site managers, foremen and operatives
- the standards prescribed
- the quality control procedures (inspection/testing) used
- the standard of workmanship by site operatives
- the definition of project priorities with speed and cost frequently outweighing the requirements for performance and quality
- the level of first-line supervision by chargehands and foremen at the workplace.

The reasons for the occurrence of such problems are themselves complex, being determined not only by technical aspects within the project but also influenced by human factors. For example, inadequacies in physical quality standards on-site could be caused by poor communication in control documentation, inadequate interpretation of the project requirements, a technical error in the design or it could result from intransigence by the workforce. The contractor's Quality Assurance Manager must, therefore, address the technical and more quantifiable aspects of quality in relation to time, cost and available resources, but must also manage the potential attitudinal problems held by the operative at the workplace. Quality Assurance, therefore, involves a socio-technical approach.

Quality assurance depends very heavily upon the Quality Assurance Manager interpreting the requirements of BS 5750: Quality Systems, and applying its principles to his role of organising, motivating and leading site management to implement the following practical measures:

During the pre-construction stage:

- Checking control documents for quality requirements.
- Reviewing the client's quality assurance system (where appropriate).
- Attending pre-contract meetings to address any aspects concerning levels of quality and its assurance.
- To review the organisation's (contractor) quality assurance system in relation to the requirements of the client and architect.

During the construction stage:

- Providing a structured and well organised approach to site organisation.
- Clearly understanding and specifying the standards of quality required to all members of the site team.
- Accurately monitoring performance and quality achieved on site (implementation of BS 8000 – Workmanship on Building Sites).
- Taking immediate action to halt work and rectify discrepancies in construction performance and quality (ensuring compliance with the requirements or Completed Product Verification under BS 5750).
- Constantly liaising with the designer to accommodate technical changes affecting quality.
- Providing formalised system for the management of material deliveries, storage, protection and distribution.
- Rewarding good quality work, rather than work speedily carried out to inferior standards.
- Engendering a sense of pride and care in the project.
- Encouraging high morale and self-achievement in craftsmanship.
- Providing skill training to meet particular requirements for quality in the project (BS 5750 requires 'quality' training at all levels within the construction industry including senior management, management

project and craft levels and extends its requirements through all sectors of the construction industry and to all professionals, in some form).

Completion, Occupation and Facilities Management

Quality assurance is equally as important in the 'final stages' of the total building process as it is in the pre-construction and construction stages. The final stage or 'post-construction' requirements for quality can be addressed in four categories:

(i) Practical completion
(ii) User occupation
(iii) Final completion
(iv) Upkeep, Maintenance and Repair

Practical Completion

At the point at which the contractor completes the building work it may be suggested that quality assurance finishes. This is far from being the case. Following the inspection of the works by the client's representative and the issuing of the Certificate of Practical Completion (a certificate marking the start of the Defects Liability Period), quality assurance should continue in the procedures adopted by the architect to first, identify and, second to obtain rectification of any deficient works. A formalised and documented procedure should form part of the initial contract to ensure that the Certificate of Making Good Defects is expeditiously obtained.

User Occupation

The occupation of the building at the hand-over by the contractor is again an area which one may think quality assurance plays a restricted role. Again quality assurance is very important and should ensure that procedures are adopted for the safety and security of the building. It should also provide for any operational manuals for equipment and services within the building and that sufficient instruction be given to 'introduce' the building to its prospective occupiers.

Final Completion

Final completion, the making good of any defects, agreement of the final account and the issue of the final certificate, should also form part of the client's initial project quality plan, ensuring as far as possible that quality assurance commences at day one of the project and continues, under his supervision to the final day of the building process in a systemised and effective manner.

Upkeep, Maintenance and Repair
In its ultimate guise, upkeep, maintenance and repair should form the final phase of the quality assurance process. In fact, as buildings become more technologically complex, include more sophisticated systems for heating and ventilation, communication and service supply, so the requirement for maintenance management or 'facilities management' is recognised.

Quality assurance should, at the completion of the total quality loop, embrace facilities management in a formalised policy of upkeep, maintenance and repair including the following aspects as its basis:

- Provision for maintenance contracts with BS 5750 certified services.
- Continuing liaison with original designers and other consultant services involved.
- Planned facilities management programme.
- Fully trained staff to execute the planned programme.
- Process for performance monitoring and feedback of the building and systems in use.
- Appreciating the ultimate need for decommissioning far into the future life of the building, a plan to provide for the safe dismantling and removal of the building fabric, structure and contents.

Taking all these aspects into the total quality loop, quality assurance embraces anything affecting quality from the client's original building notion right through to demolishing the building.

6.4 Quality Assurance: Its Application in Non-Traditional Procurement

The achievement of quality can be affected most adversely by the many issues surrounding the 'traditional' form of building procurement. As construction projects have become more technologically and managerially complex and the requirements for faster speed of production and standards of quality and performance have increased, there has been considerable emphasis given to developing fresh approaches to procurement and project organisation. These more novel, non-traditional forms of procurement aim to reduce the many problems brought about by the traditional form. They are also able to promote a different type of construction organisation in which quality assurance assumes high importance and within which project quality can be maximised. 'Design-Build' (Design-Construct), 'Management Contracting' and 'Project Management' approaches all offer alternative forms of project organisation. Design-build aims to integrate and co-ordinate more effectively the design phase with the construction process whilst management contracting and project management establish the 'management of the project' as a separate role to both the design and construction functions.

Both design-build and management contracting have considerable potential to improve the standards of quality achieved during the total building process since their procurement form promotes the following:

- Greater integration of design (and the input of other consultants) with the construction process. This allows quality assurance to become an important and integral consideration during the design and the construction phases.
- Improved communication between members of the project team allowing quality standards to be transmitted more efficiently and effectively.
- Closer co-ordination of the various construction professionals. When this occurs throughout the whole project, quality assurance becomes a commitment from inception to completion and can even progress into the commissioning and user phases of the total building process.
- The focus of responsibility to a single administrative party rather than responsibility disintegrated among various parties to the project. In essence, one party assumes full responsibility for implementing all aspects of quality assurance.
- Involvement of the contractor at the design stage. In this way performance at the construction phase can be considered at the design stage and any problems of quality achievement anticipated.
- Focus upon the team members meeting the client's genuine requirements for project performance, quality and value for money and maintaining these objectives throughout the project.

Design-Build

In the design-build form of building procurement the contractor is appointed directly by the client and becomes responsible for the whole process from initial briefing to the production of the finished building. The client deals directly with the contractor who assumes full responsibility for and co-ordinates the separate design and construction processes, including engagement of the design team. The client may alternatively employ in-house staff or separate consultants to check that the contractor is providing value for money and that performance and quality are to the standards required. This is a procurement method finding considerable support within the construction industry and one where quality is known to have been improved.

Management Contracting

The management contracting approach aims to allow the contractor to become a part of the client's team and for the total management function to be carried out in partnership with the members of the design team. The management contractor, unlike the design-build contractor, does not carry out the construction but sub-contracts the work on a competitive basis to specialist sub-contractors. Through this approach, management is established as a separate role with the management contractor directly employed by the client and controlling the project directly on his behalf.

Certification bodies have prepared Quality System Supplements to the BS 5750: Quality System standard to define the requirements for quality assurance systems in various aspects of non-traditional procurement. Lloyds Register for Quality Assurance, for example, has developed a quality system supplement for companies specifically offering management contracting services. These supplements, in addition to quality assurance standard BS 5750, form the basis for successful implementation of a quality system in the field of non-traditional building procurement.

Project Management

Whereas quality assurance can become a difficult proposition under the traditional form of building procurement, where problems arise at the various interfaces between client, designer and contractor, a single interface between the client and all other professional parties can promote the importance for the achievement of quality within a construction project. 'Project Management' is one such form of non-traditional building procurement that has the potential to achieve this requirement and has increased in popularity in recent years.

The objectives of project management are to apply management skills to the structure, organisation and control of all aspects of the construction project and optimise available resources to produce a building that better meets the client's requirements for function, cost, time and performance. With a project management structure, management is totally separated from the design process and the construction process, and acts exclusively as the client's agent in all matters concerning the project. This leaves construction professionals free to concentrate on the aspect of the project about which they know best and leaves the project management team to provide the integrating communication, co-ordination and control functions. Rather than becoming preoccupied with the many technical aspects of the project, project management can focus upon three main project aspects, those of time, cost and quality.

The services provided by a project management organisation can vary from project to project and according to each client's specific requirements. A project management organisation may carry out only a co-ordinating role under the control of the client, (Non-Executive Project Management), or can be fully responsible for the management of the total project (Executive Project Management).

The role and function of a project management organisation is to provide:

> The overall planning, control and co-ordination of a project from inception to completion aimed at meeting a client's requirements and ensuring completion on time, within cost and to the required quality standards. [1]

Project management provides the overall co-ordination of all aspects of the project on behalf of the client and is responsible for the preparation of the brief, together with the detailed programming and supervision of the design process. Project Management will also select the most appropriate form of contract and generally act as a co-ordinator during the construction process. Quality can therefore, become a continual focus of attention for management throughout the feasibility; pre-construction; construction; and commissioning stages of the project.

With singular and total responsibility for the project, project management can ensure that:

- the standard of quality required by the client is clearly identified.
- the level of quality required is reflected in the brief.
- levels of quality are clearly taken account of in the specification, documentation and drawings.
- standards of quality required are communicated to construction professionals efficiently and effectively and their various interpretations are co-ordinated.
- construction work complies with the specification and that effective quality control procedures, testing and other performance criteria are monitored.
- quality assurance continues to the commissioning, handover and use of the building through; the supervision of pre-commissioning checks and tests of equipment; the provision of operating instructions; and through monitoring the building in use during the defects liability period; and in facilities management.

Figure 6.1 shows how the quality management structure integrates the various aspects of the design and construction processes and provides constant feedback which ensures that defective work is identified early and rectified. It also provides more uniform and reliable information on project performance leading to the development of better standards and conformity to these standards.

The structure and organisation of a project management approach lends itself to the requirements of quality assurance. Project Management can ensure that only quality assured resources are procured, and furthermore develop and maintain an orientation towards quality throughout the project.

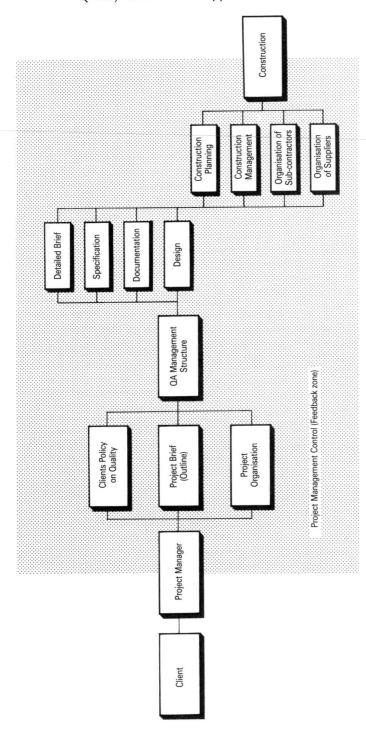

Figure 6.1 The intergrated 'Quality' flow and feedback within the project management structure

Whilst there is still a dependence upon the various construction professions providing a quality assured service to the project, there is a greater chance that their work will be better controlled and that quality will be achieved.

Quality Systems in Project Management

As the project management approach to building procurement becomes more popular in construction and project management consultant practices rapidly evolve, so certification bodies have developed guidelines for quality assurance systems with specific application to project management organisations. These systems meet the requirements of BS 5750: Quality Systems. The concept of 'total management' from inception to completion means that a quality system must meet the requirements for quality control in the procurement of design, construction, materials and other construction services.

Project management quality assurance systems differ from more general quality systems in meeting the total management concept. A project management organisation must meet all the basic criteria for policy, structure, organisation and procedures within a quality system but a number of specific requirements augment the quality system standard BS 5750.

The specific requirements should be described as follows:

Definition of Service
The service as described by BS 5750 should be interpreted as the total management of quality, time, cost and resources for a building project on behalf of the client.

Project Manager
The project management organisation should appoint a project leader, the 'Project Manager'. The project manager has authority and freedom of the client to organise, monitor and control all activities and contractual parties employed in the project. The project manager is responsible for the building process from inception to completion, including all aspects concerning the achievement of quality.

Organisation
The organisational structure should include details of the responsibility of the project manager as specified by the client, the relationship with all other professional parties, and the interfaces with the requirement for quality assurance in the project.

Design Control
The project management organisation should set up a suitable method of overseeing the design process and detail the procedure by which there is control over quality assurance in:

- the specification
- the drawings
- the schedules
- the issue of drawings and documentation
- the checking of all project information relating to design

Contract Review
There should be a contract review procedure developed which should be an on-going activity from inception to completion. The review should take account of the following requirements:

- the requirements of the client
- the design detailing of the work
- the constraints of time and cost
- the standard of performance, quality and workmanship required
- the resources needed to undertake the project
- the method and presentation of production control
- the procedures for controlling sub-contractors

Within the contract review there should be programmes of work defining the nature of a particular project and describing any requirements specific to the project. The procedures shall be fully described in the project management organisation's Quality Manual and individual projects covered by supplementary Quality Plans. All review procedures should be documented and presented at review meetings throughout the duration of the contract.

Planning
There should be a documented and planned procedure for the review of the client's need for performance and quality within the following areas:
- statutory
- technical
- health and safety
- training
- special equipment
- deliveries
- project communication and control procedures

Process Control
There should be procedures to define the method of assessing the progress of the work and the way in which performance and quality are to be

monitored and assessed. These procedures should define the actions required in relation to time scheduling and budgetary control. Provision should be included describing actions to be taken if delays and disruption occur. The method of assessing progress and reporting to the client should be clearly defined.

Purchasing

The selection of sub-contractors should be based on documented evidence of ability to perform required tasks. Selection should ideally be based upon compliance with BS 5750 but where sub-contractors do not operate a formal quality system consideration should be given to ensuring that the service they offer meets with the project management organisation's satisfaction and is included as a condition of their contract.

Where suppliers do not conform to an approved system, particular emphasis should be given to ensuring that their service is of adequate structure and organisation and that they have adequate handling, storage, inspection and test procedures for their products.

All documentation related to sub-contractors and suppliers should enable the project management organisation to trace records back to the point of origin to identify goods or services not meeting the requirements of the project and ensure they are not used on further occasions.

Inspection and Testing

Procedures should define the level of inspection of work to be carried out including the authority for accepting each stage of the work. Records validating inspection and acceptance procedures should be indexed and filed for auditing purposes.

Corrective Action

There should be procedures for the review of rejected works in order to prevent their subsequent recurrence. Records of all action taken including any resultant changes to work instructions should be indexed and filed for auditing purposes.

6.5 Quality Assurance in Building Services

Quality assurance is an increasingly growing area of interest within building services sector of the construction industry. Great strides have been made in recent years by building services organisations to develop quality assurance systems to meet the specific requirements of their construction sector, and to maintain the continuity of interest in quality assurance promoted within other construction and engineering professions. Interest in quality assurance has been sufficient to promote the formation of the

Building Engineering Services Certification Authority (BESCA), a quality assurance assessment and certification body, accredited by the NACCB.

Increasing concern for the standards of performance and quality within the building services sector has developed for a number of reasons:

- In particular building types, a substantial proportion of the construction work and building cost is associated with the provision of building services.
- As larger buildings such as hospitals, schools and office developments become more technologically complex, so the nature of building services requirements increase in complexity.
- There have been a considerable number of failures in the performance and quality of building services which can be traced to inadequacies in design, manufacture and, in particular, installation.
- As with other aspects of building construction, building services must provide a product of adequate quality for its intended purpose. In use, the quality of building services is usually interpreted in terms of performance reliability, operational cost and life expectancy of the system. In practice, it can be difficult to design and build-in 'long-term quality' to building service installations.

The Building Services Research Information Association, (BSRIA), Technical Memorandum: Quality Assurance in Building Services, [2] recognised, in the mid 1980s, that failure to meet performance and reliability standards was occurring too frequently as there was a tendency for building services systems to become more complex simply because it had become technologically possible to make them more complex. It was suggested that this tendency increased maintenance costs and incidence of failure, which could have been avoided if simpler but adequate systems were specified.

The report highlighted three principal aspects which should be prerequisite to developing effective quality assurance in building services:

- All factors that influence operating life must be recognised. Such factors will include handling, storage and installation practices together with operation and maintenance procedures.
- Equipment must be designed, manufactured and selected taking account of all the factors that will affect the performance and reliability of the finished product.
- Appropriate and adequate quality control procedures must be developed and applied to all phases of the design/installation/maintenance process as a necessary part of quality assurance.

Failure to meet the requirements specified in the client's brief may, as in all sectors of construction industry, be caused by a variety of problems arising throughout the total procurement and building process. BSRIA suggest that the most prominent reasons surround the different perceptions of requirement by the various contractual parties:

- Users expect simplicity in operation of the service, also maximum efficiency, minimum maintenance, and long-term reliability.
- Specifiers tend to over-specify but subsequent consideration to cost can often modify their views resulting in unbalanced quality levels within the same project.
- Designers tend to expand their ingenuity and ability to innovate with the result that they may present an over-complex, and over-costly design with capabilities beyond the user's real requirements.
- Installers will be mainly concerned with completion of the work in the minimum time with the maximum profit. It will be expected that all the necessary information and materials will be available the instant they are required.

Quality should be clearly specified, but more importantly, be managed and controlled such that amongst the various contractual parties there is commonality of opinion and effort.

In recent years, BSRIA and BESCA have concentrated their efforts towards heightening awareness for the principles, practices and benefits of quality assurance in the field of building services. BESCA have developed guidelines of their requirements for companies wishing to register quality assurance systems for certification. As with other sectors of construction industry, an applicant must demonstrate an ability to structure, organise and implement an effective quality system to BS 5750: Quality Systems. Procedures follow closely those already described for building contracting, although a considerable emphasis is placed upon the quality of on-site installation works, where many problems are known to exist.

BSRIA, meanwhile, have spent much time identifying those aspects of building services procurement and installation which, if addressed, could significantly improve the standards of quality assurance in building services. Important aspects include the following:

- Development of standard specifications and codes of practice, designed to set and maintain 'minimum' standards of quality.
- Adoption of 'graded' standards which specify various levels of quality, above the minimum, to allow industry to relate price to quality of service and follow the lead of the British Standards which incorporates various levels of quality in services to be selected by the user.
- Wider recognition for 'product certification' and the use of independent certification bodies to assess the standards of quality being achieved.

- Increased use of formalised procedures for supervision and inspection of quality, particularly in services installation.
- Improvement to delivery, handling and site storage of components to minimise loss and defects through damage prior to installation.
- Provision of more comprehensive information to the user in respect of operation and maintenance of building services installations.

References

(1) Chartered Institute of Building (CIOB) *Project Management in Building (1983)*
(2) Building Services Research Information Association (BSRIA) *Quality Assurance in Building Services (1983), Technical Memorandum TM 7/83*

7 Quality Assurance in Housebuilding

7.1 The National House Building Council

Housebuilding is unique in its level of commitment to formalised quality assurance. It is the only sector of UK building and construction industry that provides the purchaser with a legal guarantee of the building's performance and standard of quality, backed by insurance against deficiency and failure. This capability is provided by the National House Building Council, (NHBC), 'Buildmark' warranty scheme.

The NHBC is an independent body, financed from fees paid by individual housebuilders, and is dedicated to upholding high standards of quality in UK housebuilding. The overall aim of the NHBC is to set nationally recognised standards for building performance and quality in all new houses and to guarantee their design and construction against major deficiency or defect.

The NHBC undertakes to protect the purchaser from inadequate quality and workmanship and maintain standards throughout the housebuilding industry through implementing the following:

- Setting minimum standards of performance, quality and workmanship with which housebuilders must comply as a requirement of their continued registration as an NHBC approved housebuilder.
- Gradual improvement of the Council's minimum standards to meet the changing requirements of the housebuilding industry.
- Type approval of design and inspection of dwellings during construction.
- Provision of a ten-year warranty on the completed house under the Council's 'Buildmark' scheme.
- Impartial conciliation and arbitration of disputes between the housebuilder and the purchaser.
- Promotion of research into any aspect of housebuilding which could affect housing standards and quality.
- Dissemination of information and research to the housebuilding industry and to the public through NHBC publications.

Up to 99 per cent of all new houses built in England and Wales and around 95 per cent of new homes constructed in Scotland are approved for standards of performance and quality by the NHBC. In addition to the requirements specified by the Building Regulations and Local Authority Planning and Building Control, all new homes constructed by NHBC registered housebuilders must meet requirements set by the Council.

The philosophy of the Council demonstrated in England and Wales is the same for Scotland and Northern Ireland, although separate regional committees accommodate differences in building tradition and legal procedure.

7.2 Development of the NHBC

The formation of the NHBC, which before 1973 was known as the National House Builders Registration Council (NHBRC), was brought about in 1936, essentially to give the public some assurance of standards of performance and quality within the UK housebuilding industry.

The NHBRC gathered its membership from many professionals within architecture, building, surveying and engineering together with representatives from building societies, building industry unions and Government. Upon its formation, the Council compiled a National Register of housebuilders. Registered housebuilders were sanctioned to build and sell their houses to the standards required by the Council and provide the purchaser with a two-year warranty on the house. The Building Societies Act of 1939 gave recognition and support to the Council's housebuilders registration scheme but on the condition that insurance cover supplemented the warranty such that in the event of a builder failing in his commitment to the warranty there would be an avenue of redress for the purchaser.

By 1951, just over 40 per cent of all private sector housing was built by NHBRC registered builders, who totalled 653 in number. By 1963 these figures had changed considerably as some 1,700 registered builders constructed around one-quarter of all new private sector houses. 1965 saw major change in the NHBRC scheme as the Council extended its insurance cover on houses built by registered builders. A ten-year warranty was introduced under which the builder was required to rectify any defects occurring in the first two years following completion. After this period, the Council assumed responsibility for the occurrence of major structural defect.

A number of Parliamentary debates in 1966 resulted in a motion for compulsory registration of all UK housebuilders with the NHBRC. The Government suggested that legislation was not required as voluntary registrations had exceeded 3,000, with 40 per cent of all new private sector housing being built by registered housebuilders. Also in 1966, the

Building Societies Association gave their full support to the NHBRC and advised its members that a mortgage on a new house should only be granted if the dwelling had been built by a registered housebuilder. This was, perhaps, the most significant event in housebuilding since the formation of the NHBRC as it became virtually impossible for a purchaser to secure a mortgage on a new house which was not covered by the NHBRC ten-year warranty.

By 1968, 70 per cent of all new houses in the private sector were built by NHBRC registered housebuilders. In Scotland, however, there were only 76 housebuilders registered with the Council and, as a consequence, a Scottish Committee was formed to develop the Council's activities there. This was followed by an NHBRC Northern Ireland Committee. By 1970, 98 per cent of all private sector housebuilding in England and Wales was being undertaken by NHBRC registered housebuilders.

In using the title 'National House Builders Registration Council', it implied that the Council was a builders' organisation. To alleviate possible misapprehension, the NHBRC changed its name to become the National House Building Council, (NHBC), with effect from October 1973. During the 1970s various improvements were made to the scheme's extent and scope of cover and new specifications for standards of performance and quality introduced.

During 1981, the NHBC launched the 'Pride in the Job' campaign, aimed at raising awareness on the construction site of the need to promote quality in housebuilding. This endeavour continues to reward site supervisors who have demonstrated excellence in the supervision and management of quality through its 100 awards made annually.

A 'Conversions Warranty' was introduced in 1983 which extended the requirements applying on new houses to newly converted properties, including flats. Under this protection, the purchaser is covered by the builder for the costs of rectifying defects in the property during the first year of ownership, and by the NHBC for the cost of repair to any major defect occurring in the subsequent five years of ownership.

In 1985, under the Building Control Act, the Council became an 'Approved Inspector of Buildings'. This enabled the formation of NHBC Building Control Services Ltd, a wholly owned subsidiary, whose purpose is to provide fee paying clients with quality assurance consultancy services. A further subsidiary, 'PRC Homes Ltd' – (Precast Reinforced Concrete) was created in the same year, whose responsibility is to approve quality in repair systems used on precast reinforced concrete homes, many of which were built in the 1950s and 1960s and which now require substantial repairs.

In April 1988, the NHBC revised its warranty scheme considerably by introducing the 'Buildmark' scheme which extends the nature and scope of cover beyond its previous standards. Towards the end of the same year, the

Government widened the Council's range of activities by approving NHBC to offer a building control service to building contractors undertaking residential conversions or refurbishment of residential dwellings. Today, the NHBC's Register of Builders has expanded to well over 20 000, with virtually all new houses being constructed by registered housebuilders and an increasing proportion of conversion and refurbishment work undertaken by NHBC certified contractors.

7.3 The Maintenance of Housebuilding Standards

The NHBC is financed by fees levied on the housebuilders it registers, with the fees varying according to the builder's house prices. The finances of the Council are spent in three principal ways, on inspection of all new homes built for sale by a registered housebuilder, on providing insurance under the warranty scheme and on promoting research and information for industry and the public.

When a housebuilder applies to the Council for registration, an inspection is made of the builder's current projects to assess the levels of work and standards of quality. The builder must demonstrate all those aspects of policy, structure and organisation which accompany the operation and maintenance of a recognised quality assurance system. Many house-builders, will indeed, have quality systems meeting first- and second-party standards. Where housebuilding may be a new departure for the contractor, or where the Council is uncertain as to the builder's standards of construc-tion, the builder may be placed on the Council's Probationary Register and have its status upgraded after a minimum number of houses have been successfully completed.

Continual monitoring and appraisal of the housebuilder's work follows registration and where any deficiency is noted the case is referred to the Council's Registration Committee. Builders are always given the opportun-ity to rectify their work and increase their standards of quality before being sent formal warnings and, only in a very small proportion of cases will the builder's registration be revoked.

The principal task of the NHBC's technical staff is the inspection of houses under construction. The aim is to identify any defects which may have gone unnoticed by the contractor. In principle, each new house is visited at least once per month during construction, and in this way, many minor building faults can be detected and rectified before the dwelling is completed.

It is, however, a difficult task to decide upon the acceptable level of building performance and general standards of quality in housebuilding. The Council's main aim is, therefore, to eliminate the worst standards, rather than demand unachievable standards, and overall, to set a recognised minimum standard for performance and quality in all new houses.

7.4 The 'Buildmark' Scheme

'Buildmark' is the NHBC's sign of commitment to the pursuit of quality in housebuilding. It's symbol is a display to purchasers that their new home is protected for two years against minor deficiencies and defects and for ten years from major defects causing structural damage to their dwelling. If defects should occur, the two-year warranty ensures that the builder will carry out rectification, free of charge. If defects occur outside the builder's period of liability, the NHBC, will at their own expense, rectify the defects under the ten-year guarantee. Buildmark is also an assurance to the purchaser that the housebuilder is reputable, having been deemed competent to build by the Council, and that the house has been independently inspected for building faults during its construction.

The main objectives of the NHBC in restructuring their warranty scheme to develop 'Buildmark' were twofold, to make the scheme easier for the purchaser to understand and simplify the documentation processes for the housebuilder. Buildmark was developed in consultation with the Government's 'Plain English Campaign' and presents the scheme in a form that the purchaser can understand without being conversant with the legal aspects. Buildmark also assists the housebuilder by simplifying the registration and insurance processes. The builder now only has two major dealings with the system, first in the application to register the house for inspection and insurance, and second, in the transfer of documentation concerning the completed dwelling to his solicitor.

The Buildmark Warranty

The extent of protection afforded by the NHBC's ten-year warranty is in addition to any contractual, statutory and common law rights, that the purchaser may have against the housebuilder. The cover is 'in addition', because, under the Council's scheme the housebuilder warrants to the purchaser that the house has been built, or will be built, on the following terms:

- In accordance with all NHBC requirements for periormance and quality in construction.
- In an efficient and workmanlike manner and of suitable materials, so as to be fit for habitation.

The nature and extent of cover given under the warranty depends upon when the difficulties arise and, therefore, problems may be categorised into one of three types:

(i) Problems arising whilst the house is under construction.
(ii) Problems arising during the Initial Guarantee Period.
(iii)Problems arising during the Structural Guarantee Period.

The extent of cover, whilst the dwelling is under construction, is limited to £10 000. When problems arise, such as a builder having commenced construction but unable to complete the house due to, for example, insolvency, the NHBC will compensate the purchaser for: the amount of money above the contract price which is needed to complete the house; or any monies that the purchaser had paid to the builder and which are not recoverable from the builder.

Buildmark also provides cover for making good any defects during construction which could adversely effect performance and quality of the completed building. It is important that any rectifications are made to problems arising as they occur so that the ten-year warranty and any mortgage appraisal is not invalidated. The NHBC may procure rectification of such defects themselves or compensate the purchaser for undertaking any works.

Within the 'Initial Guarantee Period', it is the responsibility of the housebuilder to make good, within a reasonable time and at his own expense, any deficiencies, damage or defect which arise as a result of the builder not meeting the requirements for performance and quality specified by the NHBC.

During the 'Structural Guarantee Period', the housebuilder is not liable for any defects and cover is, therefore, provided by the NHBC's guarantee. This provides insurance cover for major defects considered to be within the structure of the house. Such defects may include failure of major elements of the structure such as foundations, subsidence and settlement, failures of superstructure and external envelope and fungal infestations such as dry rot. Regional building traditions are accommodated within the scope of cover. In Scotland, for example, Buildmark provides cover for defects occurring in renderings, this form of external wall finish being predominant there. Within the Structural Guarantee Period, the Council also undertakes to reimburse the homeowner for the cost of removals, alternative accommodation and storage, in the event that a major defect results in evacuation of the house whilst repair is undertaken.

Where the NHBC is liable for the cost of rectifying any defect under the Structural Guarantee Period, they reserve the right not to pay the purchaser and instead, arrange themselves for the necessary work to be carried out. This may occur in the event of specialised remedial works being required.

Under the Structural Guarantee, the NHBC are rightly aware of the basis and extent of their liability, and in a number of particular situations will not accept liability. These are as follows:

- Where any cost, loss or liability is covered by any other policy of insurance.
- Where any major shortfall in building performance or standard of quality should have been notified to the housebuilder during the Initial Guarantee Period.
- Where the item is excluded from cover by any endorsement to the basic ten-year warranty.
- Where any alteration or extension to the original dwelling may have affected material quality or expected performance.
- Where any claim for damage caused by subsidence, settlement or land heave is made by a Public Authority and the dwelling has been constructed on land which has at all times been owned by the Public Authority.
- Where any defect has resulted from the installation or presence of an unusual feature in a domestic dwelling, such as, for example a lift or swimming pool.

Outside those activities of the NHBC, little has been done within the housebuilding industry to implement quality assurance. A recent development in quality assurance implementation within housebuilding, has been the emergence of an alternative warranty scheme to that offered by the NHBC. The Municipal Mutual Insurance Company (MMI), an independent body and previously a consumer watchdog for local authority housebuilding, launched their 'Foundation 15' scheme. The scheme, as inferred in its title offers 15 years of warranty cover and operates in a similar way to NHBC's Buildmark. This opposition has, in fact, been a stimulus to NHBC. Despite the introduction of Buildmark in 1988, NHBC has already made further improvements to the scope and extent of its Buildmark scheme, and as such, remains the UK construction industry's most active and prominent supporter of quality assurance practice.

8 Quality Assurance: International Interest

8.1 European and Worldwide Dimension

In the same way that quality assurance has assumed prominence in the UK construction industry, so international interest has grown steadily, and somewhat in parallel. The basic concept of quality assurance and application of certification schemes have existed in Europe for many years, in particular within the area of product conformity for building materials and components, and to a limited extent within construction management systems. Whilst there is much commonality in quality assurance philosophy, there is some disparity in the approach, standards and systems of European countries, and even greater division in worldwide practice.

Although there is an undeniable trend towards supporting the widespread implementation of quality assurance, there remains a considerable difference in approach as building regulations, procurement systems, construction practices and attitudes towards construction differ nationally. This suggests that one generic approach to quality assurance would be unable to satisfy any global requirement. Each country has tended to develop its own approach and although there is considerable commonality of direction across Europe, following the continued expansion and harmonisation of the European Community, (EC), there is still divergence in national approach. Whilst BS 5750 and ISO 9000 have become the basis for the European Standard: EN 29000, there is considerable disparity of acceptance by member states of the European Community for this standard.

There are a number of key reasons for the distinct national quality assurance frameworks. As one might expect, differences emerge from historical traditions, and the social, economic and political demands of each country, together with the very different market forces, building methods and forms of regulation. As each country develops and refines its own systems of quality assurance and standards so there is further separation of the national requirement from any international direction.

The most discernable factors within national construction approach include the following:

- Attitudes and approach to quality in construction
- Structure and responsibility of design teams
- Organisation and responsibility of contractors
- Role of building control and policing organisations
- The nature of national markets and their perception of international markets
- Attitude and support of Government
- The nature of client organisations

It is clear that common interests in quality assurance should be reflected in the respective national quality systems and standards such that interchange in approach may take place between countries. Essentially, quality assurance serves to advertise that products, processes and services are of a high order and the more widely this can be communicated, the greater the commercial advantage for any business. The ability to trade in new or alternative construction markets, beyond merely its own national framework is an important consideration to any organisation developing a quality assurance system today.

Within the UK, all quality assurance systems are based on BS 5750: Quality Systems. This standard has considerable similarity to European standards developed by the European Committee for Standardisation, (CEN), and to the ISO 9000 standards of the International Standards Organisation and hence their adoption as EN 29000, the European Standard. The UK's national framework for quality assurance is quite similar to the national approach of France and systems used in West Germany. Such harmonisation gives considerable advantage to UK users of BS 5750 as it affords international recognition of its basic standard and allows the organisation to practice in Europe without having to invoke major change to its philosophy and organisation to accommodate the European requirement for quality. Some EC member states have said however, that EN 29000 is too biased towards UK principles of practice and disregards the national practice of other European countries, giving the UK unfair advantage.

From a UK viewpoint the adoption of BS 5750 as part of EN 29000 is likely to present both the potential for more UK companies to operate in Europe and for more European construction traffic to enter the UK. It should certainly give rise to cross-channel interchange whether that be beneficial or detrimental for the UK industry.

8.2 Problems of Integration

The greatest single obstacle to the widespread co-ordination of quality systems across European and world frontiers is the difference in national approach to building control and regulation. Although the more prominent European countries have very similar philosophies and frameworks for building control and regulation, the UK and USA construction industries serve to illustrate the great disparity that can prevail on an international level. Within the UK, all building work is subject to a national and unified system of planning approval and building control through local authority government and must adhere to one main regulatory document, the Building Regulations. In contrast, the USA has no single national building code for the USA as a whole. Instead, various standard building codes (USA building regulations), are used but are subject to a great many state and local building codes with responsibility for planning and building approval delegated to individual states and local communities rather than being nationally based. Given this situation, it is virtually impossible to implement and police any form of widespread quality assurance system, even if other factors of significance were to be unified, which indeed they are not.

The basic problem of discontinuity in quality assurance approach receives much attention, particularly in Europe. The Governments of the European Community are committed to developing standards that are internationally comparable, to remove technical trading barriers and provide a basic reference point and framework in Europe for the development of quality assurance within the single international market of the European Community. European Standard EN 29000 seeks to meet this need. It is realised however, that the implementation of this 'common standard' is futile in the immediate term as current practices differ so widely across Europe, and as such, it is more productive not to 'fully' harmonise national standards but specify minimum requirements around which member countries can develop their own standards and systems for quality assurance. This ensures that national needs are met first and foremost, but also develops an awareness of the requirement for integration on an international level. It is clear nonetheless that a longer term aim of the European Community is to develop an appropriate infrastructure across Europe within which quality assurance approval can be harmonised, systems developed and certification and testing be implemented and controlled.

8.3 International Quality Assurance and Certification Schemes

Most countries in Europe and Scandinavia have a well developed and nationally structured approach towards quality assurance. These take the form of assessment and approval systems, and in some cases full certification schemes. The framework for quality assurance in these countries differ markedly from those of North America, South America, the African continent and Asia, all of which, have far less structured frameworks at a national level. The reasons underlying this are complex and beyond the scope of this review but are due primarily to the different scale and inherent complexities of their construction industries. It would be unreasonable to directly compare the nature and activities of their industries with those of countries in Europe or the United Kingdom. Equally so, the size of construction industry, philosophy and approach across European countries should also be appreciated in this context. Different approaches to quality assurance emerge through different industry structures and whilst some European countries have close similarities others have considerable differences.

Figure 8.1 details the UK national framework for quality assurance whilst Figures 8.2 to 8.5 present four other examples of national frameworks for quality assurance, those of Sweden, West Germany, France and the USA. These represent the equivalent approach to that of the national framework for quality assurance adopted in the UK to BS 5750: Quality Systems. The frameworks of Sweden, West Germany and France represent a cross-section of quality assurance activity in Scandinavia and mid to Northern Europe. Considerable differences do exist in what might be classified as Southern European countries which include Italy, Greece and the Iberian peninsula, where construction as a whole and quality assurance specifically tends to lack the formality and control of other European countries. Southern European countries could be said to have a fluctuating approach and variable attitude due to their more unsettled economies and lacking industrial infrastructures when contrasted with say the UK, France and West Germany. The United States is included merely to give an indication of the difference in philosophy, standard and approach to quality assurance there, the US being a large construction market.

France

The national framework for quality assurance in the UK, Scandinavia and Europe, whilst maintaining similar philosophies, vary somewhat in degree of structure and level of detail in some aspects. France has, perhaps, one of the most complete quality assurance frameworks of any country, in terms of its comprehensive and precise standards with its origins dating back two centuries. France, like the United Kingdom, has adopted a national approach to quality assurance and has developed this to certification level. Co-ordinated by the Centre Technique et Scientifique du Batiment, (CSTB), {Government Technical and Scientific Centre for Construction}, France has certification schemes covering product approval, product design, quality assurance management systems, and quality assurance personnel assessment. Moreover, all aspects of construction industry are co-ordinated by CSTB through the implementation of their 'AVIS' ('technical opinion' or 'technical advice') scheme which provides certification in all construction sectors and activities.

In the field of product design, testing and approval, the Association Francaise de Normalisation, (AFNOR), {French National Standards Organisation}, awards its 'NF' Marque (Norme), equivalent in status to the BSI's Kitemark. Within the management of construction, quality systems of contractors, sub-contractors and other professionals are independently assessed (certified) by the Organisme Professionel de Qualification et de Classification du Batiment, (QPQCB) whilst design services and consultants are subject to independent assessment (certification) by the Organisme Professionel de Qualification des Ingenieurs-conseils et Bureaux d'etudes Techniques du Batiment et des Infrastructure, (OPQIBI).

Quality assurance in France is overseen by Government, in the same way that the NACCB represents the DTI in the UK framework. The Coordinating body is the Organismes Certificateur Agres, {Certification Commission for Construction}, but unlike the NACCB whose scope includes manufacturing in addition to construction, the Commission specifically serves the construction industry.

The standards of work produced in France are generally high but can be extremely variable in nature. Although formalised and comprehensive standards exist, forms of procurement or contract that are 'traditional' in the UK tend not to be used in France. Instead a greater reliance is made upon the architect/contractor relationship and the architect's contract which can lead to major difficulty and give rise to claim situations. Levels of quality can vary on a project by project basis depending upon just how well the project is designed and managed and in this respect it is subject to the same problems as any UK project.

West Germany

In West Germany, the Institut fur Bautechnik, (IfBT), {Government Institute for Building Technology}, co-ordinates and controls all planning and building approval through national policy delegated to municipalities. The certification of products is based upon Zulassung, {General Type Approval} issued by IfBT following conformity assessment of the product against state building laws and regulations. Approval of products may only follow specific requirements in design, manufacture, testing and quality control and, therefore, approval certificates are essentially the equivalent of the BSI Kitemark although their basis is in German building law. Whilst the standards of work expected and achieved is very high and national standards are comprehensive and highly regulated, it is within the area of quality management certification that the framework perhaps falls below the level of thoroughness developed under the 'AVIS' scheme in France. Quality assurance schemes covering contractors and other construction professions, although often assessed by IfBT or independent quality control organisations is, to some extent, a voluntary undertaking as, indeed, is UK certification under BSI/QAS or an independent certification scheme. Quality assurance in this area of activity is, therefore, a developing aspect at this time. It is noticeable that there is less reliance upon 'a system' making demands and that the various practices expect 'teamwork' to be the norm. Disputes for example tend to be resolved through discussion rather than by reference to contract and in general, a greater acceptance of professional and moral obligation exists towards producing a high quality value for money approach.

Sweden

Sweden follows closely the framework of West Germany although its approach is further co-ordinated through one central authority responsible to the Government with wide ranging powers. Statens Planverk, (SPV), {National Swedish Board of Physical Planning and Building}, co-ordinates and controls national policy for planning and building approval, develops Svensk Byggnorm, (SBN) {Swedish Building Regulations}, and approves products through Typogodkannande {General Type Approval}. The SPV is also responsible for overseeing Tilverkningskontroll {Production Control} which assesses quality systems used in manufacturing building materials and components. Assessment (certification) of quality management systems is undertaken by industry based quality assurance organisations, approved by SPV. Again a whole philosophy towards performance and quality exists in Sweden. Construction follows very much a Northern European System with formal procedures being used, which are highly regulated and with a much greater expectation for quality standards than perhaps would be expected in the UK at this time.

Assessment and Certification

The procedure for development, implementation and certification of quality systems in these countries follows closely the pattern in the UK under BS 5750: Quality Systems. The organisation must demonstrate unequivocally that it has developed and maintained a working quality system to the satisfaction of the regulatory standards and assessment body. The system is assessed and, where appropriate, a certificate of registration is awarded. Continued surveillance follows on a regular basis with periodic evaluation of the full system and formal re-certification. Whilst the systems used are ostensively similar the real differences tend to be the expectation and standards accepted for quality. It really is the case that some countries treat the whole concept and industry of construction with greater professionalism and regard this as a duty to the national society.

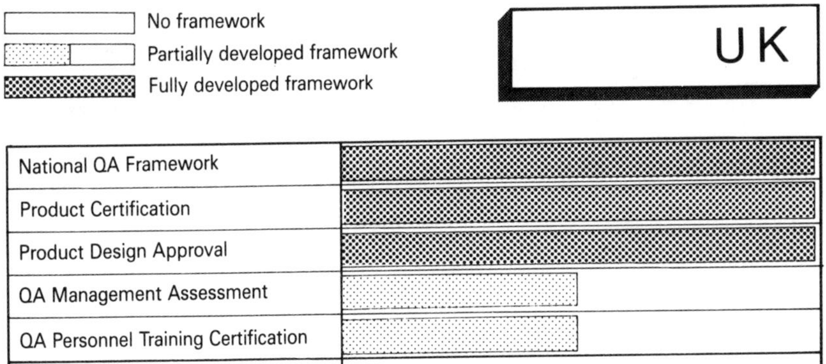

	No framework
	Partially developed framework
	Fully developed framework

U K

National QA Framework	
Product Certification	
Product Design Approval	
QA Management Assessment	
QA Personnel Training Certification	
Authorities (Planning and Building Regulation	Department of Environment: Local Authority Departments
Building Regulation	The Building Regulations
Product Approval	BSI 'Kitemark' BSI 'Safety Mark' Yarsley 'Testguard'
QA Management System Assessment	BSI/QAS Yarsley QAF Lloyds RQS ⎤⎬⎦ (Voluntary implementation)
Testing Bodies	BSI - Standards BBA - Agrément NAMAS - Testing Standard Facilities
Applicable Standards	BS 4770: Quality BS 5750: Quality Systems EN 29000
QA Co-ordination Bodies	NACCO (All industries)

Figure 8.1 The national regulation of quality assurance in the UK

☐ No framework
▨ Partially developed framework
▦ Fully developed framework

National QA Framework	▦▦▦
Product Certification	▦▦▦
Product Design Approval	▦▦▦
QA Management Assessment	▦▦▦
QA Personnel Training Certification	▨▨
Authorities (Planning and Building Regulation)	Statens Planverk (SPV) (National Swedish Board for Physical Planning and Building) Co-ordinates and controls all planning and building approval
Building Regulation	Svensk Byggnorm (SBN) (Swedish Building Regulations) Developed under SPV
Product Approval	Typogodkannande (Type Approval under SPV) Tilverkningskontroll (Production control of quality in manufacture)
QA Management System Assessment	Industry-based QA Management Systems and control organisations (Approved by SPV)
Testing bodies	Statens Provingsanstalt (Swedish National Testing Establishment)
Applicable Standards	SBN ISO 9000 series EN 29000
QA Co-ordination Bodies	SPV (Construction)

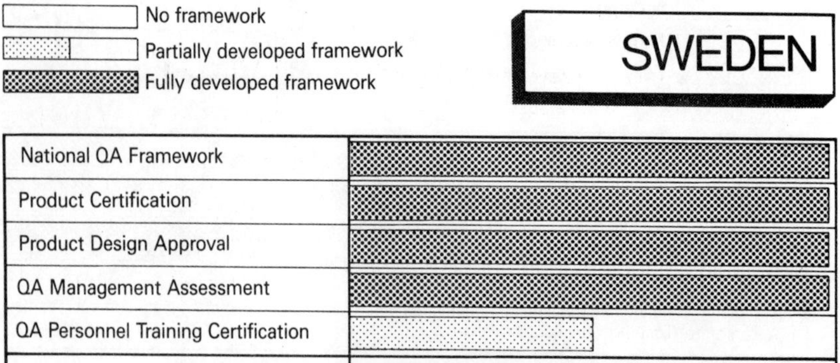

Figure 8.2 The national regulation of quality assurance in Sweden

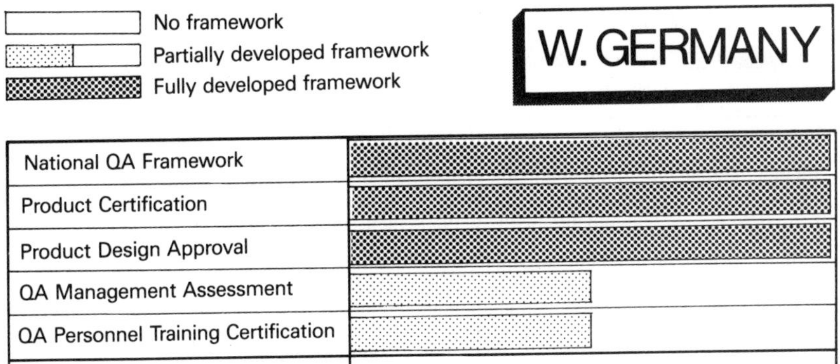

	No framework
	Partially developed framework
	Fully developed framework

W.GERMANY

National QA Framework	
Product Certification	
Product Design Approval	
QA Management Assessment	
QA Personnel Training Certification	

Authorities (Planning and Building Regulation)	Institut für Bautechnik (IfBt) (Government Institute for Building Technoloy) Co-ordinates and controls all planning and building approval
Building Regulation	Einheitliche Technische Baubestimmungen (ETB) (Uniform Building Rules) Musterbauordnung (Model Building Laws) Richtlinien (Guidelines/codes of practice) — Developed under IfBt
Product Approval	Zulassung (General and specific type approvals (IfBt)) Certificate may cover: manufacture, testing, quality control system
QA Management System Assessment	Industry Based QA Management Systems (Approved by IfBt)
Testing Bodies	Federal Material Testing Establishment (Part of IfBt and DIN)
Applicable Standards	German Standards Institute (DIN) ISO 9000 series EN 29000
QA Co-ordination Bodies	IfBt (Construction)

Figure 8.3 The national regulation of quality assurance in West Germany

No framework
Partially developed framework
Fully developed framework

FRANCE

National QA Framework	
Product Certification	
Product Design Approval	
QA Management Assessment	
QA Personnel Training Certification	
Authorities (Planning and Building Regulation)	Centre Technique et Scientifique du Batiment (CSTB) (Government Technical and Scientific Centre for Construction) Co-ordinates all construction using national approach: 'AVIS' Certified approach in all aspects of construction
Building Regulation	Building Regulations under CSTB
Product Approval	Product Approval by French National Standards Organisation (AFNOR) AFNOR Kitemark equivalent (NF)
QA Management System Assessment	CSTB – Organisme Professionel de Qualification et de Classification du Batiment (OPQCB) Assessment of construction agencies, organisations and professionals Design offices, consultant assessment scheme (OPQIDI)
Testing Bodies	AFNOR (CSTB and under AVIS scheme)
Applicable Standards	Avis Standards AFNOR Standards ISO 9000 series EN 2000
QA Co-ordination Bodies	Organismes Certificateur AGRES (Certification Commission for Construction)

Figure 8.4 The national regulation of quality assurance in France

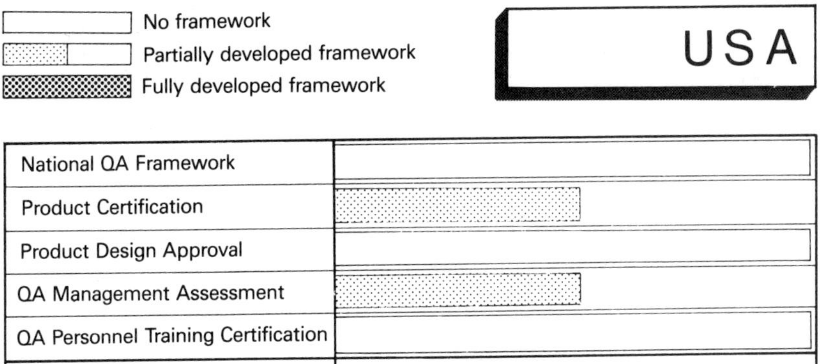

| No framework |
| Partially developed framework |
| Fully developed framework |

U S A

National QA Framework	
Product Certification	
Product Design Approval	
QA Management Assessment	
QA Personnel Training Certification	
Authorities (Planning and Building Regulation)	No Federal System of Control State Responsibility Individual states give planning and building approval, regulation delegated to local committee
Building Regulation	Standard Building Codes (Building Regulation) IBCO – Uniform Building Code BOCA – Basic Building Code SBCCI – Southern Building Code (up to 8000 local variations)
Product Approval	QA Underwriters Laboratories Inc. (Private independent product test and approval agency)
QA Management System Assessment	Individual Company Schemes (where implemented)
Testing Bodies	American Society for Testing and Materials (ASTM) (Voluntary basis) American National Standards Institute (ANSI) National Bureau of Standards (NBS)
Applicable Standards	ASTM ANSI (Not mandatory)
QA Co-ordination Bodies	None

Figure 8.5 The national regulation of quality assurance in the USA

United States of America

In complete contrast to the national frameworks of Scandinavia, Europe and the UK, the United States has no Federal Government system. Responsibility for general planning and building approval is empowered to individual states who, in turn, delegate authority to local communities. Product testing and assessment is undertaken on a purely voluntary basis by various bodies such as the American Society for Testing and Materials, (ASTM), the American National Standards Institute, (ANSI), and the National Bureau of Standards, (NBS), that is not a standards making body as such, but a research institution similar in nature to BRE in the UK.

Product approval, (certification) is available through an independent testing agency, UL Underwriters Laboratories Inc., formed by the insurance professions for specific testing procedures but which now has much wider scope. There is little formality or structure to quality assurance in design, construction and professional services and whilst larger companies involved in complex construction projects implement 'project quality assurance schemes', there is in general, no systematic approach to quality assurance since the influence of the US building codes (building regulation equivalent) are essentially very weak, adopted more in some states than others and with very disparate regulation and inspection across the US as a whole.

8.4 International Organisations Concerned with Quality Assurance

There are a number of international organisations which reflect growing interest in quality assurance and the development of co-ordinated quality standards. Some of these bodies are concerned with the development of standards themselves, product certification, and testing procedures, such as the ISO, CEN and the International Union of Testing and Research Laboratories, (RILEM). Other organisations play an important role in research and dissemination of quality assurance principles and practices, such as the International Council for Building Research, Studies and Documentation,(CIB).

The International Organisation for Standardisation, (ISO).

Since the early 1970s the International Organisation for Standardisation, (ISO), has developed over 5 000 international standards. These standards, listed in the ISO Catalogue, cover many fields of activity but the ISO standard particularly related to quality and quality assurance is the ISO 9000 series, which are modelled on BS 5750. As the ISO becomes more involved with the BSI through their membership of the International Federation for the Application of Standards, (IFAN), so BS 5750 becomes more deeply entrenched within ISO 9000. It is the consolidation of these two organisations and standards that EN 29000, the European Standard for Quality has emerged.

The work of the BSI and ISO in world quality standards through IFAN is, presently, somewhat limited to certification of products and accreditation of testing facilities. In the same way that the BSI established NAMAS in the UK, so ISO and other bodies have developed certification schemes for products and testing services in a number of countries.

The European Committee for Standardisation, (CEN)

The European Committee for Standardisation, (CEN), as the name implies, is primarily concerned with harmonising standards for quality across Europe. The Committee works closely with ISO in the development of its product and testing standards and presents their recommendations in classified documents distributed throughout the European Community. Recommendations, if appropriate, can then be incorporated into the national frameworks of the various EC member countries. The BSI is involved with the work of CEN and provides information on its activities in Europe through its bulletin 'BSI News'.

The International Union of Testing and Research Laboratories (RILEM)

The main role of the International Union of Testing and Research Laboratories, (RILEM), is the co-ordination of standards in testing facilities in Europe. Through various technical committees, RILEM is involved in a number of specific areas of interest including the field of performance and quality in the construction industry. RILEM works closely with ISO and CEN and also maintains an interest in research and dissemination through its contact with CIB.

The International Council for Building Research, Studies and Documentation (CIB)

The International Council for Building Research, Studies and Documentation (CIB), is one of the world's foremost research and dissemination organisations. The vast majority of CIB's work is through its various 'Working Commissions' of which there are around 100. Results of their work is published in technical reports many of which are used by bodies such as ISO as a guideline for the development of quality standards.

CIB has two working commissions with a specific interest in quality and quality assurance: W-86: Building Pathology, which is primarily concerned with building performance, defects and failures; and W-88: Quality Assurance, which is involved with research into quality standards, quality systems and the management of quality within the total building process.

Integration

Within the construction industry, on a worldwide basis, the role of the ISO is extremely important in co-ordinating the various national systems and standards as, indeed, are the CEN within the European Community. These bodies have only to a minor extent addressed the need for quality systems in management and concentrated upon standards for product conformity, recognising that this is the area where, in the short term, there is likely to be the greatest level of commerce between EC countries. This can be clearly seen in recent initiatives between CEN and the EC such as the draft documents outlining a European Organisation for Testing and Certification. The harmonisation of 'quality management' systems and services is clearly an endeavour for the future.

9 The Achievement of Quality Assurance: A Review of UK Research Studies

9.1 Research Interest

In recent years there have been a number of UK based research projects surrounding the concept of quality and its assurance. Some of these have focused upon performance of buildings through the quantification of building faults and defects, whilst others have qualitatively assessed the significance of design, project communication, management and supervision upon the nature and extent of problems concerning quality occurring during construction projects. Research has addressed general construction activity and specific sectors of the industry such as housebuilding, and housing studies themselves have been directed to both traditional (masonry) and non-traditional construction forms. [1,2,3,4,5]

In parallel to UK research interest in quality, there has been considerable activity in other countries. In the USA, a study of performance problems in housing has been undertaken following the identification of severe moisture related building defects in timber-framed dwellings, timber-frame being a form of construction used for many years across much of the USA. In contrast, in South America, Africa and the Far East, which, in construction terms, remain developing countries, research studies have focused upon identifying the basic problems in their respective design and construction processes which affect the quality of their national construction outputs.

Quality is, therefore, a universal interest and standards of quality achieved pose many problems at national and international level whether countries be developed or developing, or whether the construction industry be large and complex or small and simple. This chapter presents details of recent UK studies into quality to identify current and future concerns for quality assurance.

Realistically, the success of quality assurance can only be measured in terms of the 'achievement of quality'. Essentially, quality assurance implementation through QA systems must be directly linked to achieved improvement, simply to give credibility to the concept. In a construction environment in which the client will be paying for the privilege of a quality assurance system, that system must realise an unequivocal benefit. The following studies therefore, identify the achievement, or lack of achievement, of quality in building work.

9.2 Quality in General Building

Appreciating the recurring problems of defects in buildings which cost the UK construction industry millions of pounds annually, and the fact that Government is a major sponsor and client to construction, the National Economic Development Office, (NEDO), is concerned greatly with the need to improve levels of performance and quality in UK building. One of the most recent research studies addressing quality and quality assurance was undertaken for NEDO by the Building Economic Development Committee (Building EDC).

The basis of the Building EDC evaluation of quality surrounds recent investigations of communication and control of quality undertaken by the Building Research Establishment, (BRE), and presented in the report 'Achieving Quality on Building Sites' (1987) [2]. This work addresses quality achievement in a variety of non-housing construction projects, which distinguishes the work significantly from the considerable research undertaken over the last 20 years by the BRE into quality in the field of traditional and non-traditional housing.

The study primarily seeks to identify the extent to which quality failed to meet expected levels and assesses the significance of specific project aspects including management procedures, information flow, supervision and inspection in the apparent shortfalls in quality.

To appreciate the many and varied problems surrounding quality and its achievement in practice and how difficulties might, in the future, be reduced or eliminated, the BRE conducted a series of three surveys focusing upon building site quality control procedures on fifty construction sites in both the public and private sectors of industry. All the projects were essentially different and 'one-off' developments using a range of procurement and contractual procedures including traditional, design-build and management contracting approaches.

The first survey highlights how problems can arise on construction sites and the nature of deficiencies in design, organisation and workmanship that influences their occurrence. During the observation of 27 construction sites in the public sector, a total of over 500 'Quality Related Events' (QREs) are recorded. A Quality Related Event is described by the BRE as anything that causes a pause in work to consider quality during construction.

In the BRE survey, quality related events are divided into two broad aspects, within which particular causes of problems are emphasised. These including the following:

1 Aspects of design or project information causes

- lack of co-ordination of design
- design difficult to build
- unclear or missing information
- low quality design
- designer not understanding materials
- design will not work

2 Aspects of workmanship and site management
causes
- lack of skill
- lack of care
- lack of technical knowledge
- poor site organisation
- lack of protection to completed works
- poor planning by craft operatives

It was found that many of the quality related events were caused by unclear or missing project information (over 125 QREs).

Inadequacy in the quality of information itself or lack of completeness and availability are highlighted as major influences on the level of quality achieved. On a number of the sites it was thought that management spent too much time chasing late information and clarifying inadequate information, all of which consumed time which could be better spent by managing quality on site.

Inadequate workmanship was, in addition, cited as influential, though not through lack of skills but due to insufficient care in the operatives' work, (over 125 QREs). Lack of motivation is said to be a key issue in the quality achieved. The sites exhibiting desired quality were characterised by their adequate resources and trained supervisory staff who clearly understood the requirements of the project and were highly motivated toward seeking project success. The distribution of quality related events by selected aspects of the project thought to be significant is illustrated in Figure 9.1. It is also evident that once problems had occurred within these areas a satisfactory solution to the problem was achieved in most cases and only in a few cases was an inadequate solution or no solution applied.

The second survey of eleven building sites addressed private sector construction. This aspect of investigation considered if constraints of time and cost had an effect upon management's capability to resolve problems of quality once they had been identified.

It emerged that projects experiencing problems with quality were frequently projects which were behind with their programme. It was suggested that failure to meet project schedules placed management under great pressure under which the management of quality can suffer, although it is also stated that tight but realistic contract times on three of the eleven projects did not adversely affect the achievement of desired quality levels.

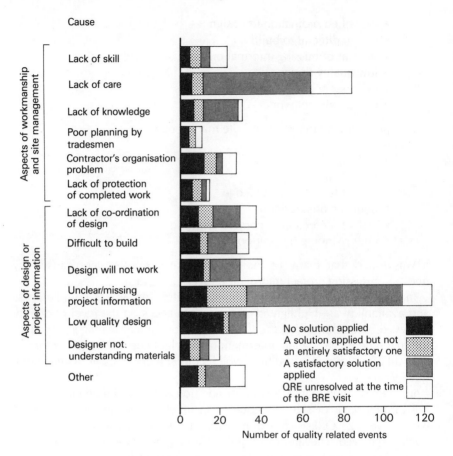

Figure 9.1 Causes of quality related events and effectiveness of resolution.
Adapted from 'Achieving quality on building sites' NEDO, (1987)

The third survey sought to examine how site staff handled problems of quality but also reviewed their roles and responsibilities under different types of contractual arrangement to see if this noticeably affected quality. It was concluded, however, that required responses to problems of quality could have been initiated equally well under any of the contractual arrangements, and in any event, the twelve sites surveyed represented too small a sample to provide tangible evidence to substantiate such a hypothesis. Whilst it was accepted that the contractual arrangements seemed to have little effect on the quality achieved, the individual management structures, which varied from site to site, were thought to be significantly influential.

The report revealed that both designers and site based managers have little difficulty in identifying problems of quality occurring within their projects but have considerable difficulty in implementing a suitable solution. Frequently, site based staff have too little authority to invoke remedial action when required and refer problems to head office staff who are, in essence, too far removed from the site to make any effective challenge to the problem.

On projects where a quality control/management supervisor was appointed, benefits were seen in the quality levels achieved, although such an appointment itself does not guarantee good quality. The quality management role was, in most cases, bestowed upon the clerk of works, the designer or another representative of the client. When quality problems arose for the clerk of works there appeared to be little back up from the architect. Quality management by the clerk of works, it was suggested, depends to a great extent upon the relationship between the clerk of works and the contractor's site manager and the level of mutual support in 'managing' quality through informal means.

Some of the more significant findings are as follows:

- Traditional specifications frequently fail to specify quality requirements for particular contracts.
- Although contractual arrangements seem to have little effect upon the quality achieved, the management structure used is influential.
- Inadequate project information is significantly influential upon achieved quality levels.
- Where deficiencies in quality occur, it is difficult to identify who is responsible for affecting the remedial action.
- Architect's representatives (clerk of works) are not always delegated sufficient responsibility to control quality effectively on site.
- Contractor's site management spend too little time 'managing' quality on site.
- Motivation is essential to the achievement of quality on site.
- Good quality relies upon good site management supervision, training and adequate resources to provide such supervision.

The report certainly highlights the 'traditional' role and actions of architects as a major source of difficulty in the achievement of quality. It criticises architects for their lack of supervision in the on-site processes, inadequate appreciation of materials and technology and the need for designers to spend more time on site. Clerks of works who frequently assume a large proportion of the work involved in quality management for the client are also highlighted, not for major inadequacies in their work, but for failure by architects in not delegating adequate authority to the clerk of works for quality management.

Contractor's site agents and construction operatives are also criticised in the report for not spending more time on quality management and lack of attention to workmanship respectively. From a wider viewpoint the report calls for fundamental changes in the current roles, duties and responsibilities of construction team members and suggests that improvements be sought to present contractual arrangements to provide a more systematic approach to quality in design and construction of buildings.

9.3 Quality in Housebuilding

Traditional (Masonry) Housing

Although quality in housebuilding is, as in other sectors of construction, not easily quantified and not always directly measurable, a number of research studies have approached the question of assessing standards of quality through the examination of 'building performance'. Quality in housebuilding depends, to a large extent, upon: the quality of the initial design, the quality of the component parts and materials specified; the quality of workmanship on site during construction; and on the occupiers' use of the dwelling. Problems arising in a completed building as a result of deficiencies in any of these aspects are problems of performance and therein indicate a possible lapse in building quality.

Poor performance and inadequate quality in housing can result from a building 'fault' in design, material or construction which manifests during the life of the building, given certain precursors, as a 'defect'. The BRE defines building faults and defects in the context of quality as follows [3]:

A building fault is:

> A departure from good practice as defined by criteria in Building Regulations, British Standards and Codes, the published recommendations of recognised authoritative bodies, and (for faults of site origin), a departure from design requirements where these were not themselves at fault.

A building defect is:

> A shortfall in performance as a result of a building fault ... whether a fault would lead to shortfall in performance (i.e. a defect) cannot be predicted with certainty but all faults have that potential.

The BRE study, 'Quality in Traditional Housing' [3], presents a clear indication of performance and quality in masonry construction, through the detailed investigation of building faults. A survey of 1 725 dwellings in both public and private sector housing in England identified over 900 individual types of building faults classified into over 100 different groups of faults. Nearly one-half (48%) of these fault types were judged to have originated on site, one-quarter originated in the design process and one-quarter attributed to deficiencies in building materials and components. See Figure 9.2.

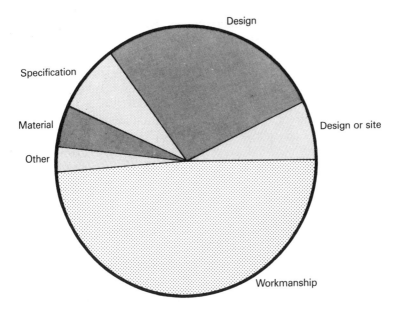

Figure 9.2 Origin of types of building faults. Adapted from 'Quality in traditional housing' BRE (1982)

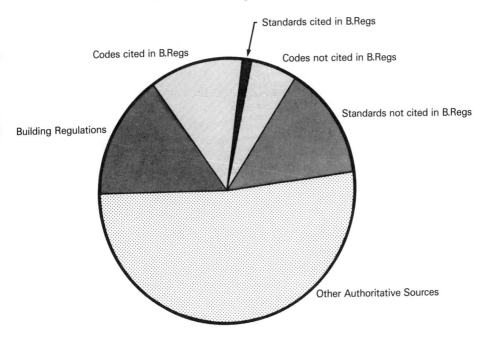

Figure 9.3 Infringement of building documentation. Adapted from 'Common defects in low-rise traditional Housing' BRE (1982)

The quarter of fault types attributed to inadequacies during the design process were thought to be direct infringements of Building Regulations or disregard for codes and standards cited in the Building Regulations. A further 20 per cent infringed codes and standards not specified in the Building Regulations and over one-half of the fault types attributed to either ignorance or conscious infringement of other authoritative documentation. The complete distribution is shown in Figure 9.3.

Many of the findings from the study of traditional housing apportions blame for inadequate quality to factors very similar to those identified in the 1987 Building EDC/BRE study. Aspects such as: inadequate information, poor communication, poor care in workmanship, and the lack of site supervision are highlighted. A most significant finding from the study surrounds the financial burden of inadequate quality. It is estimated that without improvement in design and construction quality 'houses being built now will require excessive maintenance and repair, giving rise to costs which may total up to two-thirds of the initial price'.

The study identifies that the types of problems found relate to 'traditional' and well established construction practice and therefore, accommodation of innovative practice was not in any way to blame. The largest category of problems, those resulting from poor workmanship were said to result from the general lack of care during execution of the work on site with inadequate supervision and independent inspection being a significant influence.

An important aspect of the findings highlighted that around one-third of the fault types were due to contravention of specified criteria which were unlikely to be remedied due to the high cost of remedial work once constructed. Again this identifies the clear lack of supervision and independent regulation by building control during the construction phase.

The BRE Report [3] recognises that problems surrounding inadequate quality is a major issue requiring rapid and positive improvement in quality control procedures. The need for improved communication of design, supervision on site, quality control, and skill training are all identified as significant contributors towards improving quality in the construction process.

Timber-Framed Housing

The vast majority of the UK's national housing stock, built using non-traditional construction, (forms of construction other than masonry), utilises timber-frame technology. The application of timber-frame, in the form in which it has developed in recent years, may be said to be an engineered structure and therefore, demands considerably different design criteria and site practices from those required for traditional forms of construction. It is through recognising these differences that doubts have been expressed at the performance, quality and long-term reliability of timber-framed construction. The concern for timber-frame has been encouraged, largely, by the inadequate performance and quality of other non-traditional construction used throughout the 1960s and early 1970s.

Although the total number of timber-framed houses is still relatively small in proportion to Britain's total housing stock, a large number of dwellings have been constructed in a comparatively short period of time, with little being known about the long-term performance characteristics and quality of the construction form. Technologically, 'modern', (post 1960) timber-framed housing has little in common with timber-framed construction practices preceding 1960. The performance of timber-framed housing, built before 1960, has been quite remarkable with few reports of building failure, although the performance of such older houses cannot be regarded as a reliable guide to the performance and quality of modern timber-framed housing. There have been some documented cases of structural and other problems [5], although such serious problems are believed to be isolated incidents. Recent preliminary research conducted in Scotland [1], where over 30 per cent of new houses are built using timber-framed construction, addresses current building performance and standards of quality in timber-framed housing.

Investigation identifies problems that are currently occurring, assesses their magnitude, origins and degrees of severity and discusses those factors considered to influence their manifestation. Over 100 timber-framed houses form the basis of the survey. Information was collected from two sources, first from home owners who had reported problems with their dwellings to the contractor under the NHBC ten-year warranty scheme and, second, from the detailed survey of a small number of dwellings. Whilst the information gathered cannot be defined as 'typical' since the small sample of houses cannot truly and accurately present the overall picture, the findings can be said to be 'indicative'. Certainly, the houses themselves can be said to be typical in terms of: sample of the housing stock; type of design; and construction methods used.

Specific objectives of the study were to:

- Make quantitative assessment of building faults in the sample
- Identify the nature of the faults
- Establish the origin of faults

- Assess the severity of faults and consider its defect potential
- Examine the influences upon performance and quality.

Of the 108 dwellings involved in the first aspect of the survey, 26 per cent reported problems to the contractor, 15 per cent being non-serious rectifications which could be classified as snagging items and 11 per cent representing building faults. The second aspect of the survey, the detailed survey of three dwellings identified a total of 95 faults categorised into 54 different fault types.

Examining the fault types, subjective interpretation is made as to the origin of each fault. Whereas the number of faults is quantitative, determining the origins of a fault is essentially qualitative with some faults being easier than others to interpret in this respect. Whilst some faults are clearly attributable to certain origins, others are suggested to be a combination of influences. Figure 9.4 shows the distribution of all fault types by origin.

The distribution of the origin of faults shows that 46 per cent of all faults result from site causes (workmanship) although it is likely that more than half of all faults originate from this source as an undetermined proportion of faults result from design and workmanship factors (16%) and will, therefore, be site based.

Whether a building fault will manifest as a defect cannot be predicted with a high degree of certainty. Many faults will exist quite unobtrusively until triggered by a specific influence, usually the intrusion of dampness into the structure, and, therefore, manifestation of defects is generally symptomatic of particular climatic-weathering influences. As certain infrequent events accelerate the transformation of a fault into defect some faults will have a longer incubation period than others and, in addition, the degree of severity of the fault will vary. In the survey, the faults are categorised into three groups as illustrated in Figure 9.5.

Approximately two-thirds (67%) of fault types were of a minor nature which could be said to be snagging items. Intermediate faults account for one-fifth (20%) and whilst not being of utmost concern in terms of immediate defect potential they could certainly give rise to a defect in the long term. Faults thought to be of a severe nature totalled 13 per cent. Approximately, 80 per cent of 'serious' faults were judged to have originated from deficiencies in workmanship and site practices and, therefore, could be said to be 'avoidable'.

Although some problems of performance and quality in timber-framed housebuilding may result from design such as: inadequate design detailing; inappropriate specifications of materials; failure of communication between design and the workplace, these aspects of the total building process have minor influence when contrasted with the impact of production orientated influences. The main influences upon problems surrounding quality seen in the study can be attributed to the following aspects:

- Inadequate site management to assure quality.
- Poor supervision of workmanship by first-line supervisors (foremen) and site managers.
- Poor standards of workmanship by craft operatives and general operatives.
- Lack of integration of site practices with independent quality regulatory processes (NHBC inspection and approval and local authority building control).

Although timber-frame housebuilding suffered considerably adverse publicity in the early 1980s, timber-frame is once again thriving as a construction form. The little empirical research that has been carried out on 'modern' (post 1960), timber-framed housing has shown that few major problems have been reported. Current problems concerning performance and quality in UK timber-framed housing are influenced, in the main, by poor construction and workmanship practices rather than any problems

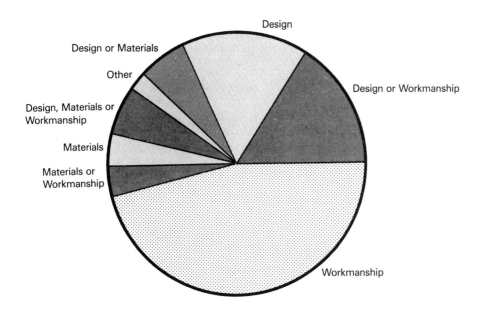

Figure 9.4 Origin of building faults in timber-framed housing

intrinsic to the design concept. Findings from studies of timber-framed housing show close similarities to the findings from investigations of performance in masonry construction with on-site workmanship giving rise to a large proportion of current problems.

Inadequate quality in timber-framed housing, as in other sectors of the construction industry, is not a problem of lacking technical knowledge and skills and not a lack of quality assurance system, in the case of housing, since the NHBC scheme provides this. Problems arise within the construction process itself which fails to integrate design and construction with the vital independent regulation by local authority building control and NHBC inspection. Problems of performance and quality in UK housebuilding clearly emerge from inadequate management of the existing systems.

As the review of research studies illustrates, the main types of problems surrounding quality in general construction, traditional and timber-framed housing, occur due to 'simple' (avoidable) errors during construction which, if more care was given to workmanship would not occur and if better supervision and regulation was invoked certainly would not go undetected.

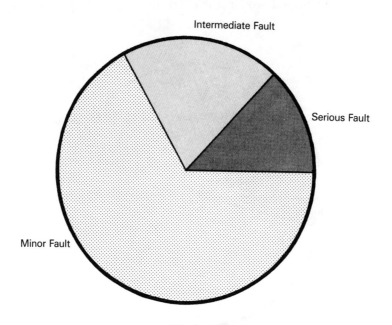

Figure 9.5 Degree of severity of faults identified

References

(1) Griffith A. *Building Performance and Quality of Timber-Framed Housing (1989) Chartered Builder Nov/Dec Edition*

(2) Building Economic Development Council (Building EDC) Achieving Quality on Building Sites (1987) *National Economic Development Office (NEDO) ISBN 0–7292–0839–7*

(3) Building Research Establishment (BRE) *Quality in Traditional Housing, Volume 1: An Investigation into Faults and Their Avoidance (1982) HMSO, ISBN 0–11–671350*

(4) Building Research Establishment (BRE) *Common Defects in Low-Rise Traditional Housing (1982) BRE, ISBN 0–85125–337–7*

(5) Freeman I.L., Butlin, R.N. and Hunt, J.H. (BRE) *Timber-framed Housing – A Technical Appraisal (1983) BRE*

Appendix 1: Definition of Terms

This appendix presents the definition of terms relevant to 'Quality' and 'Quality Assurance'. Where referenced, definitions are derived from British Standards, professional bodies and institutions, and other authoritative sources. Definitions are sequenced alphabetically, commencing with terms directly prefixed by 'Quality' and 'Quality Assurance'.

Quality
> The totality of features and characteristics of a product or service that bear upon its ability to satisfy a given need.
>
> (BS 4778)

> Fitness for purpose.
>
> (CIRIA: Report 109)

Quality Assurance
> All activities and functions concerned with the attainment of quality.
>
> (BS 4778)

> A systematic way of ensuring that organised activities happen in the way they are planned. It is a management discipline concerned with anticipating problems and creating the attributes and controls which prevent problems arising.
>
> (CIRIA: Report 109)

> An objective demonstration of the builder's ability to produce building work in a cost effective way to meet the customer's requirements.
>
> (CIOB: Quality Assurance in Building)

> A management process designed to give confidence to the client by consistently meeting stated objectives.
>
> (RICS: Quality Assurance, Introductory Guidance)

> Simple common sense written down.
>
> (CIRIA: Report 109)

Quality Audit
> The independent examination of quality to provide information. Quality auditing can relate to the quality of a product, process or system. Quality auditing is usually carried out on a periodic basis and involves the independent and systematic examination of actions that influence quality. The object is to ascertain compliance with the implementation of the quality system, programme, plan, specification, or contract requirements and where necessary, their suitability.
>
> (BS 4778)

Quality Control
> The operational techniques and activities that sustain the product or service quality to specified requirements.
>
> (BS 4778)

Quality Cost
> The cost associated with remedial work resulting from defective quality assurance practices.

Quality Documentation
> All quality assurance system manuals, procedures, plans and other recorded documented evidence to demonstrate the effective operation and maintenance of a quality system and its adherence to the organisation's quality policy.

Quality Management
> That aspect of the overall management function that determines and implements the 'Quality Policy'.

Quality Manager
> A designated senior manager who assumes organisational responsibility for the overall strategic and operational policy and administration of the quality system and who ensures that the organisation's Quality Policy is implemented and maintained.

Quality Manual
> A document, or set of documents, setting out the general quality policies, procedures and practices of an organisation.
>
> (BS 4778)

Quality Plan
> A document derived from the 'Quality Programme' setting out the specific quality practices, resources and activities relevant to a particular contract or project.
>
> (BS 5750)

Quality Policy
The overall intentions and direction of an organisation regarding quality, as formally expressed by company management.

Quality Procedure
The documented methods which describe, in detail, the principal activities involved in the quality system.

Quality Programme
A documented set of activities, resources and events serving to implement the quality system of an organisation.

(BS 4778)

Quality Review
The formal assessment, and modification where necessary, by the Quality Manager and other senior management, of the status of the quality system in implementation in relation to their set quality policy.

Quality Schedule
A document which specifies the essential items of BS 5750 which must be included in a 'Quality Plan' for a specific product or process.

Quality Standards
The recognised standards against which quality systems are assessed for conformity.

BS 5750: Quality Systems – the UK national standard for quality assurance systems.

(BS 5750)

Quality System
The organisation structure, responsibilities, activities, resources and events that together provide organised procedure and methods of implementation to ensure the capability of the organisation to meet quality requirements.

(BS 4778)

The organisation structure, responsibilities, activities, resources and events appertaining to a firm that together provide organised procedures and methods of implementation to ensure the capability of the firm to meet quality requirements established in accordance with Part 1, 2 or 3 of BS 5750.

(BS 5750)

Quality System Supplement
A document, supplementing BS 5750: Quality System, describing particular requirements of a quality assurance system.

Quality Surveillance
The systematic monitoring and verification of the use of procedures, methods, products and services and the analysis of records in relation to set standards, to ensure that the requirements for quality is being complied with.

Accreditation
The formal recognition by a national Government against published criteria, of the technical competence and impartiality of a certification body or testing laboratory.

(NACCB)

Agrement Certificate
A certificate issued by the British Board of Agrement, (BBA), formally attesting that a product has been sampled, tested and approved against set criteria for conformity, for which no British Standard yet exists.

Buildmark
The National House Building Council (NHBC) ten-year warranty scheme, backed by insurance, protecting the purchaser of a new house against deficiency and defect in performance or quality and with rectification of defects being the responsibility of the builder in the first two years of ownership and the responsibility of NHBC for the subsequent period of the warranty.

Certification
The act of licensing by a document formally attesting the fulfilment of conditions.

(BS 4778)

Certification Body
An impartial body, governmental or non-governmental, possessing the necessary competence and reliability to operate a certification scheme and in which the interests of all parties concerned with the functioning of the system are represented.

(Department of Trade and Industry)

Certification (Direct)
Where an organisation approaches an independent certification body directly for third-party assessment.

(RICS: Quality Assurance, a guidance document)

Certification (First Party Assessment)
The development of an 'uncertified' quality assurance system, approved by the client for implementation in a construction project.
Assessment)
The development and implementation of an 'uncertified' quality assurance system in consultation with the client, to fulfil particular requirements specified by the client.
Assessment)
The adoption of a full, 'independently certified' quality assurance system to the requirements of BS 5750: Quality Systems.

Codes of Practice
Document that sets down the theory and accepted principles of design practice for achieving certain aims.

(RIBA)

Conformity
Agreement between the defined and verified quality of a product or service.

(BS 4778)

Durability
The ability of an item to perform its required function under stated conditions of use and under stated conditions of preventive or corrective maintenance until a limiting state is reached.

(Harrison H.W. and Keeble E.J.:
Performance Specifications for
Whole Buildings)

Inspection
The process of measuring, examining, testing, gauging or otherwise comparing the item with the applicable requirement.

(BS 4778)

Kitemark
The registered trademark owned by BSI and only used by manufacturers licensed by BSI under a particular Kitemark scheme, indicating that BSI has independently tested samples of the product against the appropriate British Standard and confirmed that the standard has been complied with in every respect.

(BSI)

Product Approval
The declaration by a body vested with the necessary authority by means of a certificate or mark of conformity that a product is in conformity with a state of published criterial

(NACCB)

Product Standards
The acknowledged technical requirements for a manufactured product that may be incorporated into buildings in either specific or performance terms.

(RIBA)

Registration (of Quality System)
Acceptance (degree of, level of) for a quality system presented for assessment in the certification process.

Registration (un-Qualified)
Where a quality system has been assessed and no discrepancies or shortcomings have been identified.

Registration (Qualified)
Where a quality system has been assessed and minor anomalies have been identified, and which must be rectified before 'un-qualified' initial registration is awarded.

Registration (non-)
Where a quality system is assessed and found to have serious deficiency in structure, organisation or procedure, and where major amendment to the system is required.

Registration of Firms of Assessed Capability
The BSI system for the assessment of the capability of a firm to manufacture its products to specifications where there are no suitable British Standards.

(BSI)

Reliability
The ability of an item to perform a required function under stated conditions for a stated period of time.

(BS 4778)

Safety Mark

The BSI trademark awarded to products which have been tested and conform to British Standards specifically concerned with safety or to the safety requirements of standards which cover other product characteristics as well.

(BSI)

Specification

A document which prescribes in detail the requirements to which the supplies or services must conform.

(BS 4778)

Test

An inspection process in which a functional requirement is measured and observed and visually in which stress or energy is applied to the item.

(BS 4778)

Type Approval

Approval of an item based on evaluating a limited sample, on the understanding that appropriate quality assurance will be applied to the entire product of the item.

(Harrison H.W. and Keeble E.J.:
Performance Specifications for
Whole Buildings)

Workmanship

Those aspects of craftsmanship, supervision and management that combine to affect the level of quality achieved on site.

See BS 8000 – the UK standards for Workmanship on Building Sites.

Appendix 2: Sources of Further Information

Accreditation:

All certification bodies are accountable to the Department of Trade and Industry, (DTI), who is responsible for the assessment, registration and control of organisations applying for Government accreditation, through their advisory body, the National Accreditation Council for Certification Bodies, (NACCB).

National Accreditation Council for Certification Bodies, (NACCB)
80 Park Lane,
London, W1A 2BS

Association of Certification Bodies, (ACB),
The Secretariat,
British Standards Institution,
2 Park Street,
London, W1A 2BS
Association of members operating or intending to operate third-party certification schemes, recognised by the NACCB

Certification Bodies:

The following certification bodies are fully recognised by the National Accreditation Council for Certification Bodies, (NACCB), and are in their different ways connected with the construction industry. They provide certification for: product conformity; supplier and services quality management systems; and personnel involved in quality verification, within various sectors of the construction industry and within specific fields of activity.

Those references marked with an asterisk thus, *, have particular emphasis within UK construction industry in the context of this book and are described within the relevant text.

British Standards Institution/Quality Assurance Services (BSI/QAS)*
Certification and Assessment
BSI
PO Box 375
Milton Keynes MK14 6LL
   Assessment and certification of quality assurance systems.

Lloyd's Register Quality Assurance Ltd (LRQA)*
71 Fenchurch Street
London EC3M 4BS
   Independent third-party certification of quality assurance systems.

Yarsley Quality Assured Firms Ltd (YQAF)*
Yarsley Technical Centre
Trowers Way
Redhill
Surrey RH1 2JN
   Independent third-party certification of quality assurance systems.

British Approvals Service for Electric Cables (BASEC)
P.O. Box 390
Breckland
Linford Wood
Milton Keynes MK14 6LN
   Certification of products and supplier quality systems in its specialised
   field.

Ceramic Industry Certification Scheme Ltd (CICS)
Queens Road
Penkhull
Stoke-on-Trent ST4 7LQ
   Independent third-party certification of products and quality systems
   within the ceramics industry.

UK Certification Authority for Reinforcing Steels (CARES)
Oak House
Tubs Hill
Sevenoaks
Kent TN13 1BL
   Independent third-party certification of products in its specified field.

Product Assessment and Certification Organisations:

A number of institutions and organisations within the field of construction industry provide, essentially, product conformity services. These are listed below. where a listing is repeated from appended 2.2 Certification Bodies, this indicates that the organisation concerned has a specialist division for product conformity certification.

British Board of Agrement (BBA)
P.O. Box 195
Bucknalls Lane
Garston
Watford
Herts WD2 7NG
> Testing, assessment and certification of products and processes used in construction.

British Standards Institution (BSI)/Testing Services
Test Centre
BSI
Maylands Avenue
Hemel Hempstead HP2 4SQ
> Assessment of products under the registered trade marks 'Kitemark' and 'Safety Mark' schemes.

Yarsley Quality Assured Firms Ltd (YQAF)
Yarsley Technical Centre
Trowers Way
Redhill
Surrey RH1 2JN
> Product conformity certification under its 'Testguard' scheme.

Lloyd's Register Quality Assurance Ltd (LRQA)
71 Fenchurch Street
London EC3M 4BS
> Product conformity certification.

Other Quality Assurance Related Organisations:

The following organisations are involved in the promotion of quality assurance within the UK construction industry, some within the general field of building, whilst others have specific areas of interest.

Building Research Establishment (BRE)
Garston
Watford
Herts WD2 7JR

An Executive Agency within the Department of the Environment, promotes research and publication in all aspects of construction, including considerable works relating to quality and quality assurance.

Building Services Research and Information Association (BSRIA)
Old Bracknell Lane
West Bracknell RG12 4AH

Promotes quality assurance research and publications and with BESCA have developed a quality systems certification scheme with building services sector of industry.

Chartered Institute of Building
Englemere
Kings Ride
Ascot
Berks SL5 8BJ

Promotes quality assurance in construction through its Professional Practice Board, published 'Quality Assurance in Building' (1988) through its specialist Quality Assurance Working Party and Quality Assurance in the Building Process (1989).

Construction Industry Research and Information Association (CIRIA)
6 Storey's Gate
London SW1P 3AU

Research and publication in all fields of construction has major interest in quality assurance, published CIRIA Report 109: Quality Assurance in Civil Engineering (1985).

Department of the Environment (DoE)
Construction Industry Directorate
Romney House
43 Marsham Street
London SW1 3YP

Relates Governmental policy concerning construction and, in specific fields such as quality, quality standards, and so on.

Department of Trade and Industry (DTI)
Standards and Quality Policy Unit
Ashdown House

123 Victoria Street
London SW1E 6RB
> Responsible for Government policy on quality assurance standards, including the construction industry sector.

National House Building Council (NHBC)
58 Portland Place
London W1N 4BU
> Responsible for NHBC 'Buildmark' ten-year warranty scheme, promotes all aspects of quality assurance in housebuilding. Described in detail in Chapter 7.

National Measurement Accreditation Service (NAMAS)
National Physical Laboratory
Teddington TW11 0LW
> Joint activity of British Calibration Service (BCS) and National Testing Laboratory Accreditation Scheme (NATLAS), quality testing of products.

International Quality Assurance Organisations:

The following institutions and organisations have interests in quality and quality standards and promote quality assurance on an international level. Some of the bodies integrate research activity including the field of quality assurance and as such may provide further useful sources of information. These references are marked with double asterisks thus, **, within the listing.

Commission of the European Communities (EC)
Bruxelles
Belgium
> Co-ordination of quality assurance certification schemes throughout EC countries.

European Committee for Standardisation (CEN)
Bruxelles
Belgium
> Co-ordination of standardisation throughout EC countries.

European Union of Agrement (UEA)
Paris
France
> Product approval co-ordination of Agrement system in EC.

International Council for Building Research, Studies and Documentation, (CIB)**
Rotterdam
Netherlands
Worldwide organisation promoting research in all fields of construction, including a working party specialising in quality assurance.

International Organisation for Standardisation (ISO)
Geneva
Switzerland
Responsible for co-ordination of worldwide standards, including those relevant to quality systems.

International Union of Testing and Research Laboratories (RILEM)
Paris
France
Co-ordination in EC of testing systems, facilities and standards in relation to products used in construction.

Recent Publications

Quality Management in Construction – Certification of Product Quality and Quality Management Systems (CIRIA Special Publication 72, 1989)

Quality Assurance – Guidelines for the Interpretation of BS 5750 for use by Quantity Surveying Practices and Certification Bodies (RICS, 1990)

BS 8000 – Workmanship on Building Sites (BSI, 1990)

Quality Management in Construction – Interpretations of BS 5750 for the Construction Industry (CIRIA, 1990)

Index

accreditation 7, 24
  definition 25, 135
  body 7, 17, 25
  process 7
  certification bodies 7, 25, 139
  see also National Accreditation
  Council for Certification Bodies
  (NACCB) 7, 25, 139
agrement
  Agrement Board 27, 47, 141
  assessment 27
  certificates 47, 135
American Society for Testing and
  Materials (ASTM) 115, 116
American National Standards
  Institute (ANSI) 115, 116
architect
  client's selection of 73
  responsibility for quality 69
  role in quality assurance 69, 76
AVIS quality assurance scheme
  (France) 108, 114

BBA (British Board of Agrement) 27,
  47, 141
British Standards (BS) 27, 28
  in construction 27, 28, 81
British Standards Institution (BSI)
  27, 28
  Kitemark scheme 7, 35
Safety Mark scheme 35
BS 4778: Quality 13
BS 5750: Quality Assurance 14
BSI/QAS (British Standards

Institution/Quality Assurance
  Services) 7, 26, 36
Building Economic Development
  Committee (Building EDC) 120
Building Engineering Services
  Certification Authority (BESCA) 67,
  95
building regulations 81
building services 94
Building Services Research and
  Information Association (BSRIA) 95,
  142
Buildmark 98, 102

certification 24, 34, 135
  bodies 17, 24, 34, 139
  definition 25
  process 25, 36
  schemes 25, 34
Chartered Institute of Building
  (CIOB) 142
checklists 56
clients 73
  role 69, 73
codes of practice
completion 86
conformity 136
construction
  application of quality assurance
   in 3, 5
  awareness of quality within 1, 66
Construction Industry Research and
  Information Association (CIRIA) 1,
  1, 142

consultant's role 76
contractor 66
   client's selection of 73, 77
   responsibility for quality 77
   role 77, 82
cost planning 42

defects
   in buildings 124
   in timber-framed housing
      126;current building performance
      and quality of timber-framed
      housing 127
   in traditional housing 124; quality
      in traditional housing
      124; common defects in low-rise
      traditional housing 125
definitions 133
   Department of Environment
      (DoE) 142
Department of Trade and Industry
   (DTI) 7, 8 142
   represented by National
   Accreditation Council for
Certification Bodies (NACCB) 7,
   139
design
BS 5750 76, 78
design-build 88
faults and defects 124
designers
   responsibility for quality 76, 80
   role 76
detailing for manuals 54, 58
durability 137

economics 11, 42
engineering industry 4
   European Committee for
Standardisation (CEN) 106, 143
European Community (EC) 103
European standards 103
   EN 29000 103
expectation 16

first party assessment 33
fitness for purpose 3
formalisation 2, 5, 17
framework for quality assurance 3,
   50
France 108
   quality assurance in 108, 114
   approach 108
   organisation 108, 109
French Government Technical and
   Scientific Centre for Construction
   (CSTB) 108
French National Standards
   Organisation (AFNOR) 109
function 16
functional requirements 16

guidelines for development of
   quality assurance systems 6

housebuilding 9, 98
   construction 124
   faults in 124
   defects in 124
   insurance 9, 98
   non-traditional 126
   quality assurance in 101, 124
   traditional 124

independent assessment 7, 34
initial registration 37, 137
inspection in quality systems 36
insurance 9, 98
International Council for Building
   Research Studies and
   Documentation (CIB) 117, 144
International Organisations for
   Standardisation (ISO) 116, 144
international quality assurance
   schemes 107
International Union for Testing and
   Research Laboratories (RILEM) 117,
   144

kitemark 7, 35

laboratories 27
  materials and testing 27, 28
life of buildings 16
Lloyd's Register Quality Assurance
  (LRQA) 7, 34, 40, 140, 141

main contractor 77, 82
maintenance 86
marketing 10
materials
  certification 27
  handling 80
  procurement 77, 81
  testing 27, 80, 81
  use 85
management 52, 66
  of construction 66, 69
  design 76
  site 82
  total building process 2, 66
manuals for quality systems 54, 58
Mutual Municipal Insurance (MMI)
  9, 104

National Accreditation Council for
  Certification Bodies (NACCB) 7, 25
  139
national framework for quality
  assurance in:
  France 108
  Sweden 110
  UK 3, 111
  USA 116
  West Germany 109
National House Building Council
  (NHBC) 9, 98, 143
National Measurement
  Accreditation Service (NAMAS) 27,
  143
National Testing Laboratory
  Accreditation Scheme (NATLAS) 27

non-registration 37, 137

objectives 2
  of quality assurance 69
  organisational 50
organisations
  quality assurance 139, 141
  international 143
outline design 70, 76

performance
  building 124
  housing 9
  specifications 28
personnel assessment 26
philosophy 21
  organisational 21, 49
practical completion 86
products 26
  approval 26
  certification 35
  conformity 26
  standards 27, 28
procurement 2, 69
  design-build 88
  management contracting 88
  project management 89
  traditional 69
  sub-contractors 77, 80, 82
project management 89
  quality systems 92

quality
  assessment 32
  audit 31, 61, 133
  control 56, 133
  cost 11, 133
  definitions 13, 132
  documentation 31, 50, 133
  in building , 17
  in construction 5
  in engineering 4
  in housing 9
  manager 52, 133

management 14, 52, 133
manual 31, 133
plan 56, 133
policy 31, 49, 55, 134
procedure 31, 55, 134
programme 31, 134
review 31, 57, 134
schedule 57, 134
standards 23, 28, 134
surveillance 31, 135
systems 54, 134
system supplement 30, 135

records 31
registered firms of assessed
  capability scheme 27
registration 37
  non-registration 37
  of quality systems 36, 37
  qualified 37
  unqualified 37, 137
reliability 137
research organisations 13, 9117,
  144
Royal Institution of Chartered
  Surveyors (RICS) 67

safety mark 35
schedules 57, 134
second organisations 57
second party registration 33
services 94
site management 82
  supervision 83
specifications 138
standards 23, 28, 134
standardisation 107, 116, 144

aims of 107, 108
  organisations 143
  international organisations 143
surveillance 31, 135
Sweden 110
systems 49

terms used 132
test 138
  definition 138
testing 27
  accreditation 24
  bodies 117, 144
  laboratories 27
third-party registration 34
timber-framed housing 126
traditional housing 124
training 57
type approval 98, 109

unqualified registration 37, 137
upkeep 86
USA 115, 116
user requirements 71, 74, 86

value 21

West Germany 109
  general type approval 109
  Government Institute for Building
    Technology (IfBt) 109
  systems/schemes 1109, 113
  standards 113
worldwide interest 105

Yarsley Quality Assured Firms Ltd
  (YQAF) 7, 34, 39, 40, 140, 141